Howard W. Sams & Company

Complete Guide to Video

Howard W. Sams & Company

Complete Guide to Video

By John J. Adams

PROMPT®
PUBLICATIONS

©1997 by Howard W. Sams & Company

PROMPT© **Publications** is an imprint of Howard W. Sams & Company, A Bell Atlantic Company, 2647 Waterfront Parkway, E. Dr., Indianapolis, IN 46214-2041.

All rights reserved. No part of this book shall be reproduced, stored in a retrieval system, or transmitted by any means, electronic, mechanical, photocopying, recording, or otherwise, without written permission from the publisher. No patent liability is assumed with respect to the use of the information contained herein. While every precaution has been taken in the preparation of this book, the author, the publisher or seller assumes no responsibility for errors or omissions. Neither is any liability assumed for damages resulting from the use of information contained herein.

International Standard Book Number: 0-7906-1123-6
Library of Congress Catalog Card Number: 97-68179

Acquisitions Editor: Candace M. Hall
Editor: Natalie F. Harris
Assistant Editors: Pat Brady, Loretta Leisure
Typesetting: Natalie Harris
Indexing: Natalie Harris
Cover Design: Debra Delk
Graphics: John Adams
Graphics Conversion: Jeremy Adams, Deshane Roberts, Christina Smith, Kelly Ternet, Terry Varvel
Additional Text: Jim Adams
Additional Illustrations and Other Materials: Courtesy of Sega of America, Sharp Electronics Corporation, Sony Computer Entertainment, Sony of Canada Ltd.

Trademark Acknowledgments:
All product illustrations, product names and logos are trademarks of their respective manufacturers. All terms in this book that are known or suspected to be trademarks or services have been appropriately capitalized. PROMPT© Publications, Howard W. Sams & Company, and Bell Atlantic cannot attest to the accuracy of this information. Use of an illustration, term or logo in this book should not be regarded as affecting the validity of any trademark or service mark.

PRINTED IN THE UNITED STATES OF AMERICA

9 8 7 6 5 4 3 2 1

CONTENTS

Preface   1

CHAPTER 1
Introduction to Home Video Basics   5
   Home Entertainment Centers   7
      Mini Home Entertainment Centers   13
   25 Video Terms You Should Know   15
   Summary   23

CHAPTER 2
Sights, Sounds & Signals: All About TVs   25
   The Definition of Television   26
   How Television Broadcasts Work   26
   Ancient TV Technology   26
   The Three "S"s   27
   Sight   28
      The Picture Tube   28
      How Images Are Created
          On The Screen   30
      Components Of A Modern Picture Tube   33
      Improved TV Tube Technology   34
      Alternate TV Picture Technology   35
      Considerations When Choosing
          A Picture Tube   38
      Features Explained   41
   Sound   45
      The Sound Experience   45
      Sound Options   46
      Sound Terms   46

What Constitutes An Average Home Theater Sound System?	51
Equipment	51
Final Sound Bites	52
Signal	52
EM Waves	54
Modulation/Demodulation And Carrier Waves	54
What Makes Up A TV Signal	55
Air And Cable	58
Future Mediums And Signals	58
Final Wave	58
Interface And Control	58
Remote Controls and Menu Options	58
Types Of Remotes	59
On-Screen Programming And Control	60
Wrap Up	61

CHAPTER 3
Receiving With Antennas, Satellite Dishes & Cable — 63

Antennas	63
Terms	64
Antenna Types	65
Installation Notes	67
Cable TV	67
Satellite Dishes	71
C-Band Satellite Equipment	76
DBS	77
Future Of Satellites	83
Summary	84

CHAPTER 4
VCRs, Laser Disk Players & Digital Video Disk Players — 85

Video Cassette Recorders (VCRs)	86
Basic Components Under The Hood	87
Other Circuits	90
Types of VCRs	92
Editing With A VCR	93

What To Look For In A VCR:	
Recommendations	*93*
VCR Repairs	*94*
A Final Note On VCRs	*95*
Laser Video Disk	95
Digital Video Disk Players (DVD)	98
Wrap Up	101

CHAPTER 5
Camcorder Basics — 103

The Essential Camcorder	104
Formats	*104*
A Camcorder's VCR Components	*107*
CCDs, MOS and CMOS	*107*
Lenses	*108*
Viewfinder	*110*
Features	*111*
Sound	*112*
Digital Camcorders	113
The Digital Difference!	*113*
Camcorder Accessories	117
Summary	118

CHAPTER 6
Video & Audio Editing — 119

Editing Principles	119
Types Of Editing	*120*
Methods Of Editing	*121*
Linear vs. Nonlinear Editing	124
Title Makers, Video/Sound Effect Mixers & Video Processors	*126*
Putting A Package Together	*127*
Expanding Your Linear System	*128*
Expanding Your Nonlinear System	*128*
Steps To Linear Editing	*129*
Steps To Nonlinear Editing	*130*
A Few Tips On Making Home Movies	130
Wrap Up	130

CHAPTER 7
Video Game & Internet Consoles — 131

- Video Game Consoles — 132
  - *Terminology And Definitions* — *132*
  - *Basics* — *133*
  - *Console Advances* — *137*
  - *Considerations* — *138*
  - *Games* — *138*
  - *Where To Get Game Cartridges And CD-ROMs* — *141*
- Console Emulation And Copiers — 141
- Hooking Up Your System — 142
  - *Troubleshooting* — *143*
- Internet Consoles — 144
  - *Technology* — *145*
  - *Internet Appliances Currently Available* — *147*
- Network Computers — 149
- The Future — 150

CHAPTER 8
Purchasing Help & Recommendations — 153

- Consuming vs. Purchasing — 153
- Know The Jargon — 154
- Research Before You Buy — 154
- Purchasing Fraud — 155
- Where Should You Buy Video Equipment? — 156
- Where You Should NOT Buy Video Equipment — 157
- The Best Time To Make A Purchase — 157
- Purchasing Strategies — 157
  - *Other Purchasing Tips* — *158*
- Purchasing Recommendations — 159
  - *Improving Your System* — *159*
  - *Television Set* — *161*
  - *A/V Home Theater Receiver* — *163*
  - *VCR* — *163*
  - *Camcorder* — *164*
  - *DBS* — *164*
- Rating A Brand — 165
- Wrap Up — 165

CHAPTER 9
Features — 167

Television	167
Sound	168
Home Theater A/V Receivers	169
Remote Control And On-Screen Programming	170
Direct Broadcast Satellites (DBS)	172
Video Cassette Recorders (VCRs)	172
DVD	175
Camcorders	175

CHAPTER 10
Basic Building Blocks Of Video — 179

Hooking Up Your Home Entertainment System	179
Connector And Cable Basics	180
The Purpose Of Wiring	*180*
Connectors	*180*
Connector And Cable Types	*181*
Notes On Cables And Connectors	*187*
Other Basic Building Blocks Of Video Systems	188
Planning Your Connections	190
Basic Connection Rules	*191*
Know Your Basic Building Blocks Of Video	191
Hooking Up An Antenna	*192*
Hookup Diagrams	196
RF Audio	*196*
Hooking Up A Television With A Band Splitter	*196*
Hooking Up A Television And VCR With A Band Splitter	*197*
Hooking Up A Television And VCR With An Internal Band Splitter	*198*
Hooking Up A Cable Decoder	*198*
Hooking Up A Cable Signal/Decoder Signal Selector	*198*
Hooking Up A Cable/Antenna Signal Selector	*199*
Hooking Up A Mono VCR	*200*

Hooking Up A Stereo Hi-Fi VCR	*201*
Hooking Up Two VCRs To Copy Tapes	*201*
Hooking Up A Direct Broadcast Satellite:	
One LNB/IRD	*201*
Hooking Up A Direct Broadcast Satellite:	
Two LNBs/IRDs	*202*
Hooking Up A Laser Disk Player	*202*
Neat Hookup Gadgets	203
The Future	204

CHAPTER 11
Troubleshooting Video Equipment — 205

It Is Usually Something Simple	205
Know When To Call In The Pros	206
Common Problems	207
Troubleshooting Guide	209

APPENDIX A
Web Addresses — 215

APPENDIX B
Product Manufacturers — 219

Index — 221

To Kristy ...

... for her life, love and creative energy.

@}-'-,-'-,——

PREFACE

The TV has become the largest icon of modern society. The future really arrived for us when we were able to instantly view scenes from around the globe. From there came the ability to record these new glowing images. Now each of us is able to take a camcorder and cast actors in our own screenplays. It is almost unimaginable to think of a time when there were no televisions or VCRs. It's hard enough to conceive of a house without at least three TV sets and two VCRs beaming signals of electrons directly into our brains.

This form of media continues to grow, churning out nightly news programs, movies, sitcoms, video games and video interaction. What is the equipment brings it to us? TVs and monitors glue our eyes to video entertainment, and siblings and spouses fight for remote control dominance. VCRs accept plastic black cassettes full of films and copyright notices. Camcorders etch in stone those priceless memories for all to see, including the ones we wish to forget. Video game consoles let us relax with a quiet game of video chess, or get frustrated at being demolished by the enemy for the thirtieth time just before we reach our target. Interactive consoles with built-in CD players bring us sights and sounds from a digital world, then bring us back to reality when a 5-year-old kid helps to set it all up.

In today's typical home, there is a main home entertainment center, as well as other mini-entertainment centers throughout the house. There are "normal" components such as TVs, VCRs, integrated stereo systems and camcorders. There are newer devices such as digital small satellite dishes, video game systems, digital video disk players and TV-to-the-Internet boxes. It all comes together to form 10%-25% of our waking time, day or night.

How does a TV work? How can you improve your satellite dish's reception? Which brand of VCR should you purchase? What features should you look for? How do you put it all together? Video technology often seems so transparent that we never ask these questions. This prevents us from finding out just how new equipment works or how to improve our existing equipment.

Complete Guide to Video, the book in your hands, explains video systems in an easy-to-understand language. It outlines the common components of modern video/audio equipment, and gives details and features of the newest gadgets. Hopefully this book will succeed in revealing to you some of the mystery of the wires spaghetti behind your modern-day components. Maybe it will even leave you with knowledge that will help you conquer more advanced systems in the future.

Videophiles will also be interested in learning of the newest updates to equipment. Features that were previously only realized on video equipment costing thousands and thousands of dollars are now available for a mere fraction of the cost.

Complete Guide to Video begins in Chapter 1 with an introduction to the typical home entertainment center and mini-centers, a brief explanation of the newest technological wonders, and 25 video terms you should know. On to chapters 2 through 4, to give you all the information you need to know about TVs, antenna basics, cable TV, DBS and standard satellites, VCRs, and the newest digital video disk technology. Chapters 5 and 6 cover camcorders and video processing basics.

Chapter 7 briefs you on current video game consoles and Internet devices currently available to you and your family. Chapter 8 gives you information on how to plan what you will need and how best to purchase video equipment. Chapter 9 expands on this by giving you a list of the features to look for and the ones to avoid. Chapters 10 and 11 explain, in easy-to-understand language, how to pull the system together and troubleshoot any problems.

Complete Guide to Video will help with your shopping choices and answer the questions you may have asked salespersons from whom you never received a *true answer*. Once you have this information in hand, you will be able to buy your video equipment with confidence. No more mystery in society's current largest commonality: *video technology*.

There were a lot of people who helped with the creation of this book. Thanks to Kristy for her love, tolerance and superior (tough!) editing abilities. To my family for their help, support and love: Pam Cassa, Jim, Jason, Jennifer and Jordan Adams. Special thanks to Jim Adams for helping in the writing and compilation of this book. To Missy for her long hours of critiquing, direction and typing help. To Dorothy Christiansen for her friendship and writing inspiration. To Mike Jackson for guidance and friendship. To Gordon McComb and Ian Masters, the giants on whose shoulders I stood. To Candace Drake Hall, Natalie Harris, and all of the PROMPT Publications staff for their expertise and kind words.

Enjoy!

John Adams

CHAPTER 1

INTRODUCTION TO HOME VIDEO BASICS

"If it weren't for Philo T. Farnsworth, inventor of television, we'd still be eating frozen radio dinners." Johnny Carson

Video has become completely interlaced into modern civilization's fabric. It is a technological weave upon which we base our lives, loves, personalities and clothing. What makes up this video pattern? What does the average home video center look like these days?

In a typical home, three TV sets blare the sights and sounds of the times. See Photo 1-1. Children's eyes are fixated on a purple dinosaur teaching them basic morals. The minds of teenagers are being fed a constant stream of adult drama they hope to live up to. Movies bring our fantasies to life, if only for a few hours. Millions of alien wars are fought and won as video game systems become nannies to our children. We are movie producers and directors with camcorders and videotape archives our lives. We can

Photo 1-1. A 27" stereo TV on a stand. Reproduced with the permission of Sony of Canada Ltd.

Chapter 1: Introduction to Home Video Basics

order cable or satellite TV, and discover whole new worlds of video shows. The list goes on.

Is this the average use of video technology? What uses are in your home, and what uses would you like there to be? Maybe if you install that new miniature satellite dish, you can pick up another few hundred shows and (unfortunately at a one-to-one ratio) infomercials. However, before investing in a system, if you break up video and learn to explain its parts, you can better understand it and make better choices for your personal video system.

Video is derived from a Latin word meaning I SEE. Video is commonly thought of as only TV and its various components. In the context of this book, video means a combination of sight- and sound-producing electronic equipment and components: in other words, those which stimulate the eyes and ears of the viewers, immersing them in a visual, tonal

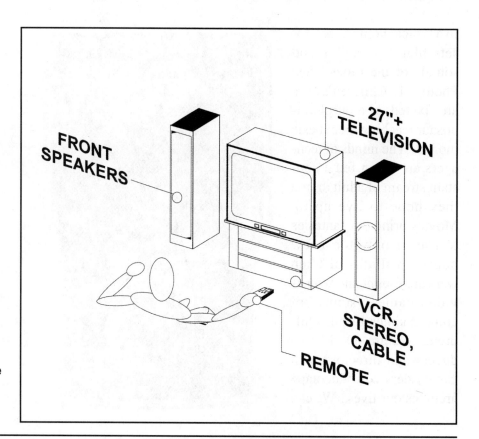

Figure 1-1. The average home entertainment center.

and emotional experience. This includes the tons of support electronics that go into bringing this sensory feast.

HOME ENTERTAINMENT CENTERS

What is a home entertainment center? Back in the 1950s and '60s, an ordinary home had one television around which the whole family huddled. There was no fighting over what to watch, as there was only a handful of TV shows. Video recording devices were nonexistent. You either watched a show live or missed a part of television history. The rabbit ears on top of the TV had to be adjusted continually. Sometimes a good picture appeared when you formed a human chain and everyone lifted their left leg. Home movies took weeks to develop; then you had to fight with the screen and projector, and coax unwilling family and friends to sit and suffer through vacation films.

The definition of *home entertainment center* changes each time a new technology is released, but let's pin down the basics:

With a home entertainment center, we bring the visual and audio experience of a movie theater into our living rooms: walls of equipment, miles of wire and a comfy chair in which to sit. In a perfect video world, this setup is fine. In reality, our expectations need to be more down to earth (and on a budget). Let's look at an average home theater, an expanded system, and a light system. If you get stuck on a few of the video terms, please see the glossary in the back, or review the section called *25 Video Terms You Should Know*.

Refer to Figure 1-1, which shows an average home entertainment center. Figure 1-2 displays a light system, and Figure 1-3 shows an expanded home entertainment center.

The TV Set. The boob tube. The idiot box. The telly! It's the most fundamental building block of your home entertainment center. Don't make a stingy choice with this one: it will last you 7 to 15 years. Who wants to be reminded of a bad purchasing decision for that long?

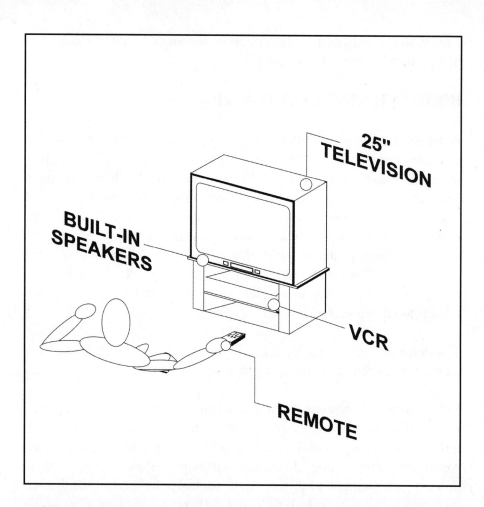

Figure 1-2. A "light" home entertainment center.

Average: A salesperson would have you believe that a 60" projection TV is the only way to go. Not true! Whatever TV set is already in your living room is probably plenty. Typically, a 27" is used in a small- to mid-sized room. Invest most of your video dollars into screen size, quality and reliability now, and worry about sound after the next few paychecks come. Try to hold back on fancy (but useless) features like picture-in-picture (PIP).

Expanded: If you want to immerse yourself in the visual experience and have space and cash to spare, go for a 31" - 35" tube model TV. Look for as many helpful features as your dollar will allow. A 40" to 60"+ wide screen or projection TV would definitely be an upper-end choice.

Light: The absolute smallest TV you get if you intend to build up your home theater is a 25" tube. Try to stay away from mass-market brands such as GE, unless you aren't picky about picture quality. You will get almost no features, but this is a good start.

Recommendations: Features are one of the biggest cost raisers of TV sets. Dumping the PIP feature alone can save you over $100. Put your money into a bigger, higher resolution screen and anything that will increase the quality of the picture itself, such as comb filters. Another thing to consider is that digital TV (DTV) will soon be upon us, making older low-resolution sets obsolete. Look for a higher-resolution set now rather than risk having to buy one in three or four years. Don't worry if the unit doesn't have a built-in stereo or surround sound feature; you can always add these features later. Most importantly, always get a remote control in the bargain.

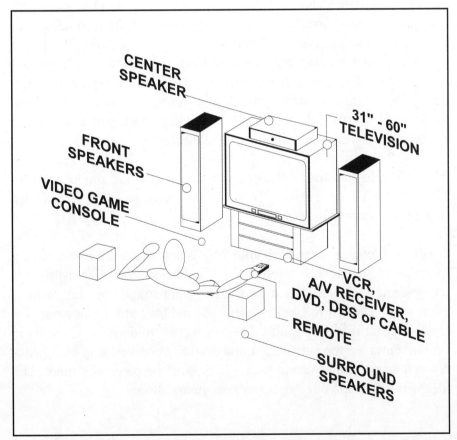

Figure 1-3. An expanded home entertainment center.

Chapter 1: Introduction to Home Video Basics

Sound Systems. Next to sight, sound is the human sense that gets the most stimulation from modern home video systems. It starts with headphones that produce sweet sounds, all the way to full digital reproductions in digital surround sound. A home theater is not complete without a fair- to high-quality sound system. You don't necessarily need $10,000 Dolby Digital Surround with THX and all the bells and whistles to enjoy an award-winning soundtrack; a simple stereo unit will be just fine. One step better would be surround sound of some form.

Average: This ingredient has received heavy transfusions of technology in recent years. Sound quality is on par with current movie theater sound systems. In some cases, they are thousands of times better. While purchasing a TV set, try to find a unit with built-in surround sound, or at least with an MTS stereo receiver, and a built-in amp with speakers.

Expanded: The ultimate sound experience is a surround sound receiver/decoder/amplifier. This will totally immerse your mind in the sound experience. Most electronics stores offer a nicely packaged home theater surround sound system. This usually includes a Dolby Pro-Logic surround sound decoder/receiver, a built-in or external amplifier, and five speakers. The most recent development is in digital surround sound: Dolby Digital, also known as AC-3. If you want an all-out system, go for one with a THX Cinema package that will provide you with a sound system equal to or greater than a movie theater.

Light: The lowest sound choices you should take are mono speakers built into the TV, or at least a TV set with an audio jack that you can use to plug into your home stereo.

Recommendations: Start with your TV. When you order a new set, try to get one with external stereo jacks. Plug these into your existing stereo system. Always work toward a surround sound package, with at minimum a Dolby Pro Logic system. Spend the extra money now for Dolby Digital if at all possible (it saves having to do it in two or three years). Purchase the surround sound decoder/receiver, amp and speakers as a home theater sound package, or buy one piece at a time. The idea is to build up your system as your funds allow.

VCR. VCRs have become as common a video component as televisions. Their biggest attraction is the fact that most movie rental stores only offer the VCR format for prerecorded feature movies. Look for a hi-fi stereo, four-head unit. It's well worth the few bucks to do away with monotonous mono units. The winning feature of VCRs is their ability to record. This is not currently possible with other prerecorded media technologies.

Average: A hi-fi, four-head, remote-controlled VHS model with on-screen programming.

Expanded: Although Super-VHS is a far superior product, steer away from it unless you are doing professional-quality video editing. I have yet to see a movie store that rents out S-VHS tapes. Look for a four-head, hi-fi unit with front A/V jacks, on-screen programming via remote, VCRPlus+, cable readiness and plug-and-play. If you will be editing, go for the models with features that are geared toward this, such as a jog/shuttle.

Light: For $150 or less, you can find a run-of-the-mill two-head mono sound unit. If there is absolutely no room in the budget for a hi-fi VCR, then don't get one.

Recommendations: Try to change your mind if you are considering a mono VCR. Spend the extra $50 now if you can. If you have trouble setting the clock and channels, get a plug-and-play VCR; they will do the work for you. Last but not least, find a unit with VCRPlus+ and save yourself some programming grief.

DVD Player, or Laser Disk Player. Digital video disk players are now a viable alternative to laser disks: high-resolution images and Dolby Digital all on a disk the size of a standard audio CD. Impressive! Laser disk will soon go the way of Beta, 8-track and Disco. If you have a laser disk unit now, don't despair. Laser disk is still a high-quality way to view prerecorded movies.

NOTE: Neither DVD nor laser disk are recordable yet.

Recommendations: Invest in DVD now if you can. If not, wait until a recordable unit is available and the prices drop. If you have a large screen television, I highly recommend a DVD over a VCR for prerecorded movies. The picture quality is surprisingly superior to a low-resolution VCR signal.

Cable Boxes, Digital Satellite Receivers & Antennas. You need to receive a broadcast video signal in order to watch TV. Many new options have exploded onto the market in recent years. Cable is the old standby, but direct broadcast satellite (DBS) services are giving the cable companies a run for their money. A good antenna perched on your roof is always a good investment (especially with cable's occasional unreliability) and in the case of DBS, a necessity: you cannot watch local channels without it.

Recommendations: I am going to leave this one up to you. Cable is the old standby, but the new DBS antennas are becoming a VERY popular, and really are a superior quality alternative.

Camcorders (Plus Accessories). Like the old adage goes, "What came first, the chicken or the egg?" A video needs to be turned into an electronic signal before it is transferred across the world. The video camera does this for anything you wish to point it at and record. Now anyone with $400 and a keen eye for movies can whip out near-broadcast quality home movies.

Recommendations: If you just want something for recording fond memories, a basic VHS-C or 8 mm model will do. If you are doing high-quality productions such as weddings or videotapes to sell, then a Hi-8 or S-VHS-C and support equipment is a must. The higher quality of these units is unbelievable. At the top-end of camcorder technology are the new digital units. However, I recommend waiting on the $2500-$4000 investment in digital units until they are more integrated with computer standards, and cheaper.

Video Game Consoles, Internet Appliances and other Computing Options. Don't think that these things are just for kids. Adults are just as addicted to this "drug" of the video world as children. The processing power and memory of these units are pushing light speed. There are even ways to access the Internet with them. See Chapter 7 for the various options on this type of entertainment.

The Viewer. The most overlooked factor in the home entertainment equation is the person. You should treat yourself as the most fundamental part of a home entertainment outfit. You need a system that fits your tastes. Think of it as a nice new set of clothes. Do you want flashy, classic, comfortable or simple? Design a system according to YOUR tastes. Don't forget the purple easy chair with built-in 'fridge and a remote.

MINI HOME ENTERTAINMENT CENTERS

Today most homes have multiple mini entertainment centers. 99% of Americans own a TV, 67% own two or more, 95% own a VCR, and 30% own a camcorder. Gone are the days when there was only one big TV set in the house.

A Story. Here is an exaggerated example of a mini home entertainment center. Walking into your home, you hear a scream coming from a distant room. Gun shots make you dive for cover behind a sofa in the family room. With great courage, you peek over the couch to see a huge 60" projection TV. At the side you see a set of massive speakers. Reality sets in as you see Mel Gibson's face and hear yet another explosion echo from the black towers in the corner.

You notice that at the side of the massive TV cabinet is a rack full of audio equipment. Lights are dancing to the music coming from the movie. On the top of the TV is a box with a numeric display and plastic card sticking out: the satellite receiver.

You decide you have seen *Lethal Weapon* too many times, so you slide into the kitchen for a bite to eat. A woman is crying. After one last turn down the hallway you see a 19" TV set on the kitchen counter playing today's episode of *All My Children*.

There's a quick reheat of day-old pizza and a quick catch-up on who's marrying who on the soap opera. With pizza in hand and a quick grab of the mail, you head for your bedroom. As you're walking, you can hear the kids screaming at each other: "You died!" "My turn." "Gimme it, Billy!" A quick peek into the kids room reveals a fight for supremacy

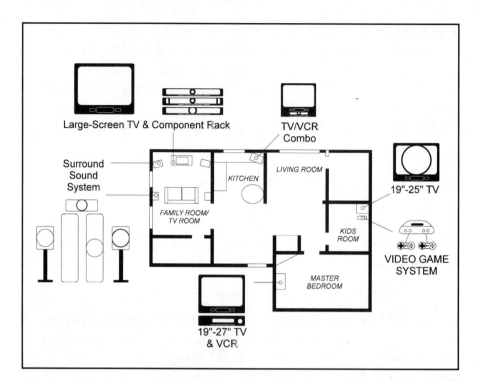

Figure 1-4. Mini-entertainment centers throughout the home.

of the game pad. On their 19" TV, you see images from a video game called Mortal Kombat. No use in trying to break it up. "You guys forgot to turn off the TVs!" you say. Only mumbles as the fight goes on.

To your bedroom for a short rest. On the bed sits your spouse, operating a remote for a 25" TV with a VCR attached to it, playing a home movie that was made on the family's camcorder last summer.

You sit on the edge of the bed and open your mail: the electric, cable and satellite bills! You pass out from the number of digits in the tally box.

Conclusion. From the previous example, you can see there are many small entertainment centers throughout the home (in addition to the main one). The home entertainment center's purpose has not changed since the days of early storytelling radio and TV. What has changed is that most homes now have multiple video centers planted in various rooms.

Each person in a household has different viewing tastes, requiring multiple TV sets and support equipment throughout the home.

Look at Figure 1-4 to see the layout of an average home and its main home entertainment center, along with several mini centers. Following this layout pattern, you will be able to better set up your own systems and choose the correct equipment for the correct room. No over or underpurchasing!

25 VIDEO TERMS YOU SHOULD KNOW

Here are a few tidbits of tech talk to tide over your video curiosities. Subsequent chapters will contain more detailed definitions and explanations.

Analog/Digital: See Figure 1-5. A video signal's information is either analog or digital. What does this mean? Analog information is represented by continuous and smoothly varying signal amplitude (height of a wave) or frequency (number of times it alternates per second). Sound

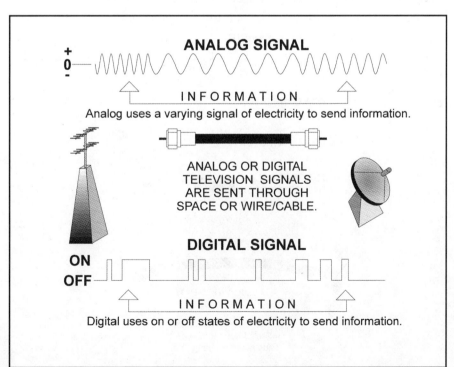

Figure 1-5. Analog and digital signals are used throughout video.

is an example of an analog signal. Digital implies a state of either on (1) or off (0). Information is represented by these two states as opposed to being continually varied. By using analog-to-digital signal converters (ADC), a digital microprocessor can handle the signal information in a much more efficient manner. The signal is then converted back to analog by a digital-to-analog converter (DAC) and piped out to speakers or TV screens, etc.

Antenna: A device which pulls the broadcast signals from the air for use by our TVs. The rods on the antenna each pick up different channels and culminate them into one cable. The information is then piped to a piece of video equipment which translates it to picture and sound.

Audio/Video Connectors or Jacks: Also known as A/V lines, RCA jacks or cinch connectors. These are the connectors and lines that route TV signals between your equipment. They carry the composite video signal and its companion, the audio line. Both terminate with an RCA-type of connector. The reason multiple lines are used instead of a single coaxial cable is because the signal is less prone to degradation caused by the continual combining and uncombining from one component to another. Refer to Figure 1-6 to see the various connectors and cables used in home entertainment systems.

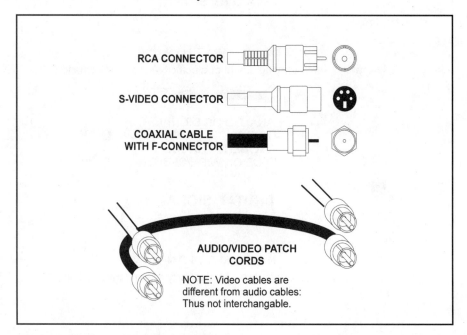

Figure 1-6. Typical video connectors, cords and cables.

Broadcast Signal: The TV signal coming to you from a local TV station, cable company or satellite service. Currently, only analog broadcast signals are transmitted.

Cable TV: Programming which is broadcast to our home via coaxial cable. A local cable company receives various movies, shows and events from satellites, divides them up, and sends them to our hungry eyes and ears for a price.

Camcorder: A combination video camera and videocassette recorder. The reason this definition is here is to remind you that at one time these units were entirely separate and bulky; hardly portable. In fact, some units are now called *palmcorders* because they fit snugly into one hand.

Composite Video Signal: When a TV signal is sent to our home it is not in a format conducive to video components. It needs to be separated into video and audio signals (baseband). The video signal is called a composite video signal. It is used to route the signal to the various video components of your entertainment center, via the audio in/out or video in/out jacks, for example.

DBS/DSS, or Direct Broadcast Satellite and Digital Satellite System: Sometimes used as a generic term for small satellite systems, but DSS is actually a trademark owned by Hughes Electronics Corp. In the past, a large satellite dish was needed to receive a fairly weak signal from a satellite. Now, with new high-power digital satellites, you can receive hundreds of channels with a tiny 18" diameter dish. DBS is challenging the cable companies to become the top subscriber service. For around $400 you can get the satellite equipment, and for as little as $10 a month you can subscribe to basic services.

Diagonal Measurement and Aspect Ratio: Refer to Figure 1-7. TV picture tubes are measured diagonally; top left corner to bottom right corner. The ratio of the screen's width to height is called the aspect ratio. Most televisions on the market are a 4:3 aspect ratio. The newer HDTV and DTV formats may allow for a 16:9 widescreen ratio which is conducive to the format most movies are filmed with.

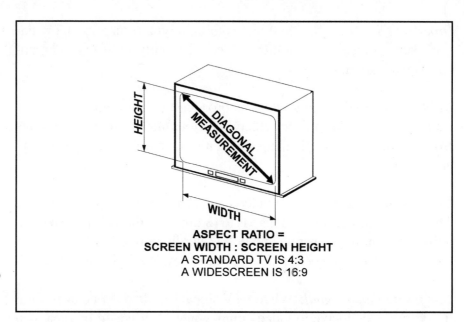

Figure 1-7. Diagonal measurement and aspect ratio of a television screen.

Dolby: A compression and expansion noise reduction system developed by Ray Dolby. Nearly all consumer electronic sound devices use this technology to improve the quality of the signal and to reduce the amount of noise in the system. The Dolby trademark is owned by Dolby Laboratories, Inc. It is merely a technology licensed to manufacturers, not a product.

DTV and HDTV: These are two distinctly different terms that are often seen together. Digital television (DTV) is a new digital broadcast standard. It makes use of new compression techniques and digital wizardry. The goal is to bring in much better resolution and sound, and (if broadcasters have it their way) six times the channels. This will totally replace the NTSC standard by the year 2006. This will eventually make it necessary for you to replace your current NTSC TV with a new DTV set, or at least require you to add a converter to your existing set. High definition television (HDTV), on the other hand, is a signal format which provides extremely high-resolution images and high-fidelity sound. It is not currently available in the U.S. See Chapter 2 for further DTV/HDTV information.

DVD, Digital Video Disk: The DVD format is purely digital. It has the highest video resolution of any video product (except HDTV) and has digital surround sound capabilities. It hooks up to your home entertainment system and plays either a full-length feature film or an audio CD. DVD can fit a whole movie onto an audio-sized CD. Currently these are only playback units but soon there will be machines with recording capabilities. Good-bye, VCR!

Hi-Fi, High Fidelity: This describes a sound equipment's ability to record or play back a near exact reproduction of a wide range of sound frequencies with little distortion. What this means is that your hi-fi equipment allows a pristine recording and playback of sound.

Home Entertainment Center: A combination of consumer video and audio equipment placed into a central area of the home. This would include but is not limited to a TV, VCR/laser disk player or DVD, antenna, cable box or satellite receiver, a stereo or home theater sound receiver and amplifier, speakers, camcorder, video game consoles, Internet appliances and miles of support wiring. Don't forget the comfortable chairs.

Home Theater Decoder/Receiver/Amplifier: Home theater equipment is usually sold as a combination: Dolby Surround decoder, receiver and built-in amplifier. There are three basic options: Dolby Surround, Dolby Pro-Logic and Dolby Digital. If you can afford it, buy the building blocks separately and go for a THX option. This will replace the stereo components in the video equation. Who wants to listen to a movie's award-winning sound effects or music track with runty, underpowered speakers?

Internet Appliance. This is a new category for home entertainment. The growth and popularity of the Internet has spawned a new device which allows you to access the World Wide Web and E-mail through your TV. Currently, TV top units are for sale which attach to your TV and allow you to "surf" via remote control. In the future, they are likely to be built into TVs and become an integral part of video and our society in general, much like the television itself.

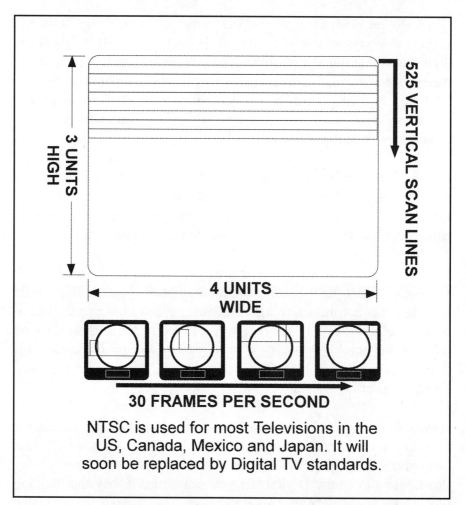

Figure 1-8. The NTSC signal. (National Television Standards Committee)

NTSC (National Television Standards Committee): Refer to Figure 1-8. This is the current color TV broadcasting standard used in the United States, Canada, Mexico and Japan. It is a signal that converts to 525 lines of resolution at 30 frames per second, and is broadcast from every TV station in this country. Most TVs currently sold in North America comply with this standard. You will often see the term *NTSC* on new computers and Internet appliances: it merely means they can output their signals to a standard TV. PAL (Phase Alternate Line) is a standard used in European color TV, which features 625 lines per frame and 25 frames per second.

Resolution: This is a typically a generic term used to indicate the sharpness or vividness of a television's picture. Technically, it indicates the number of vertically placed horizontal lines used to make up an image on your TV screen. (See Chapter 2 for a more detailed explanation.)

S-Input: Also called S or S-VHS connect. A new type of 4-pin connector used in modern video equipment. It separates the video signal into a form that is more conducive to newer electronics.

Satellite: A man-made object that orbits the earth. It sends and receives TV signals in a much higher frequency range to compensate for natural interference and the distances involved. A satellite dish or satellite antenna is a ground-based unit which picks up TV signals from this satellite.

Signal: Information such as video, sound, timing, etc. Signals take the form of electromagnetic waves or impulses transmitted through space or wires. In video and audio, the information (signal) can be in either analog or digital form.

Stereo: The recording and playback of two channels of audio: two microphones for recording and two speakers for playback. This allows for a realistic sound reproduction that our ears hear naturally.

Surround Sound: The 3D of sound. Surround sound encodes multiple channels then decodes them and plays them back through multiple speakers. A minimal surround system has four speakers. Two in the front are in stereo, and they also simulate a phantom center "speaker." The two at the sides of your head pump out a mono surround channel. The next level of surround sound has five speakers and one subwoofer. The phantom center of the previous system is replaced with an actual speaker usually placed on top of the TV in front of the listener to hear dialogue from the movie. The third surround option, and the newest, uses one additional subwoofer to add that punch to your movie's soundtrack. In addition, the surround speakers are also in stereo. See Chapter 2 for more surround sound explanations and setup tips.

HOW VIDEO CHANGED OUR LIVES

A jet fires its engines in preparation for takeoff. Location: a naval carrier a hemisphere away. Across the world, a man is preparing the most important speech of his life. Can he convince the American people to go to war?

A stealth fighter armed with a laser-guided smart bomb and a kamikaze camera transmits ghostly images of buildings drawing closer and closer until their inevitable destruction.

"Just two hours ago, Allied air forces began an attack on military targets in Iraq and Kuwait. These attacks continue as I speak. Ground forces are not engaged." President George Bush, January 16, 1991.

The Gulf War represented the *ULTIMATE* use of video technology (as well as defense technology). CNN broke the first news of the erupting conflict. For months camera crews were on stakeout, waiting for the clock to tick off the last presidentially allowable seconds. The night sky filled with the light as news crews sought cover under the nearest objects. However, in the true spirit of the news, the cameras still rolled and visions of war were transmitted to satellites hovering thousands of miles above the earth. U.S. citizens become transfixed to their TV sets. Spellbound at the coverage, they didn't realize they were seeing history in real time. No other generation has had this technological option.

This marked the beginning of a new form of war: a war of *up-to-the-second, 24-hour* reporting of events that were happening half a world away. Never before had we seen images of fighting delivered to our homes as they happened. The ultimate in TV technology arrived when generals and modern aircraft were more famous than actors.

Not only did we see live footage of our aircraft delivering ordinance to the enemy, but we were also privy to the other end of the barrel, so to speak, all due to modern video equipment. Cameras transmitted the CNN signals to awaiting satellites. The satellites streamed their bits and waves down to our neighborhoods. Cable boxes routed the signals to our homes. TVs unscrambled the jumbles of waves into images for our curious eyes to view. VCRs recorded the images for spouses at work. How long did this go on? Was it a mere one-hour newscast? Hardly! It was 24-7 when the last shot was fired and the last smart bomb was released from an F-117 Stealth. (Not to mention all the video games that came out at the time with a war-theme, or the entertainment and educational programs on the Gulf War aircraft, etc.)

So, if anyone asks you, "What is the ultimate test of TV technology?", tell them a *LIVE-coverage war*, such as the Gulf War.

Video Game Console. Any of the many personal video game units such as the Nintendo NES/SNES/N-64, Sega Genesis/Saturn, Sony Playstation. Their games are stored on a video game cartridge or CD-ROM.

Video Signal: The electromagnetic wave used to transmit video and sound information. This can mean a signal transmitted to your home via antenna, cable or satellite. It may also be used to describe the routing of these signals within your home entertainment system via A/V cables. See Chapter 2 for a further explanation.

SUMMARY

The face of video technology and its equipment is ever changing, with brighter, more vivid images, sound systems that rival the top theaters, and more ways to receive quality programming. By keeping up with the technology and learning about new terms and features, you will make smarter purchasing decisions, saving your money and sanity. Now let's take a more detailed look at these video components.

CHAPTER 2

SIGHTS, SOUNDS & SIGNALS: ALL ABOUT TVs

"TV: Chewing gum for the eyes." Frank Lloyd Wright

The television set is the nucleus of a video system. It is the most important element of any system next to you, the viewer, of course. If you desire, you can expand from your TV's core a video system to be proud of for years.

Your most intelligent video buying decisions are required to purchase your TV set or monitor. Don't be nervous. You won't need to be a nuclear physicist to develop scientific shopping skills. All you need is the knowledge and lexicon of TVs: how they work, how one set differs

Photo 2-1. A Sharp 32J-S400 32" television. Reproduced with the permission of Sharp Electronics Corporation.

from others, what features should have top priority, and which features should be reconsidered for fear of bankruptcy.

In this section you will learn the language you need to decode a salesperson's spiel. Never again will high-tech video terms give you an inferiority complex.

THE DEFINITION OF TELEVISION

The origin of the word *television* best describes its meaning. First we have the Greek word *tele* meaning *far off*. Then we have an Indo-European word, *vision*, meaning *to see*. The technology of TV now allows us to view anything from our living room.

Refer to Photo 2-1. Television is a combination of systems. There is the picture tube and support circuits, the speakers and sound circuits, and the signal circuitry which decodes the sights and sounds sent to us from ever-increasing distances around the globe. Together these make up a modern television set.

HOW TELEVISION BROADCASTS WORK

Inside a video camera is an electronic gizmo called a CCD (charged coupled device). It picks up light that is being bounced off surrounding objects. The light is transformed into an electronic signal and pumped out to our homes live or recorded for later broadcast. Either way, the electronic signal is transmitted through the air or via a wire/cable. The television in your home then steps in by grabbing the signal and translating it back into light. Thus the signal it allows you to view a "light show" from virtually anywhere in the world. See Chapter 5 for more on CCDs.

ANCIENT TV TECHNOLOGY

Television's technology is still a prisoner of World War II. In 1939, the first commercial television was sprung onto the world, a toy compared to modern sets. The war then put a halt to the expansion of TV technology. Shortly after the war ended, a standard was established and TVs were popping up in every neighborhood in America. The low-quality NTSC video signal standard, used to deliver Ed Sullivan to the homes of baby boomers, became securely entrenched in the future of the video industry. The enemy, consumer complacency, has remained after all these years. After all, we are used to a low quality image, at least until now: the new DTV standards have arrived.

Sure, the picture quality has improved over the years, and the size of the screen has reached backbreaking size, but resolution has fallen out of cadence. (Resolution is a measurement of the amount of detail in a picture tube.) This chapter will help explain why the TV sets of today are nearly the same technology as those of the 1950s and 1960s. More importantly, it will explain the newest standards and improvements and answer the question, "Should I buy a TV set with this new technology (HDTV or DTV)?"

THE THREE "S"s

Divide a TV set into its components and you get S*IGHT, SOUND* and *SIGNAL*. This is why it is called a television *set*. To a lesser degree, you may add *interface* and *control*. At any rate, with the three subsystems, you can give your mind a daily fix of sensory input through two senses, to be exact: auditory and visual. The picture tube gives you visual stimulation, and the sound system and speakers give acoustic stimulation.

How do you improve your video system? By raising the standards of your TV set's sight, sound and signal technology. First of all, your TV is a balance of these three elements. See Figure 2-1. An improved TV picture tube would give brighter, crisper, more visually stunning images. A surround sound decoder would immerse your ears in sweet sound. Improved signal quality would in turn increase the quality of both sight and sound by delivering more vivid imagery and more defined sounds.

To better understand and improve each of the "S"s, let's look at the workings of each. Also, we'll check out the breakthroughs that are making their way to your local video equipment dealers.

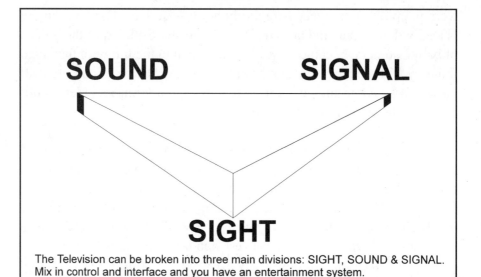

Figure 2-1.
The three "S"s of television.

SIGHT

THE PICTURE TUBE

Your window to today's world takes the form of an ancient invention called a *cathode ray tube*, or *CRT* for short. This is the actual *screen* upon which the panoramic vistas of your favorite movies are electronically painted. Hundreds of thousands of dots are dispersed across the screen to form streams of pictures before your eyes. Let's look at how they work and what features are available.

Color. Refer to Figure 2-2. By looking carefully through a magnifying glass at the picture tube, you will see the fine lines and miniaturized dots which make up an image. These are commonly called *pixels*. An average 28" TV screen has about 450,000 pixels.

Pick one pixel from a pure white area of the screen. Look closely, and you will see that the square or circle is composed of three small bars or circles of color: red bar on the left, green in the middle and blue on the right. This is called a triad. Up close you can see three color bars; but as you move back, the appearance becomes that of a white dot.

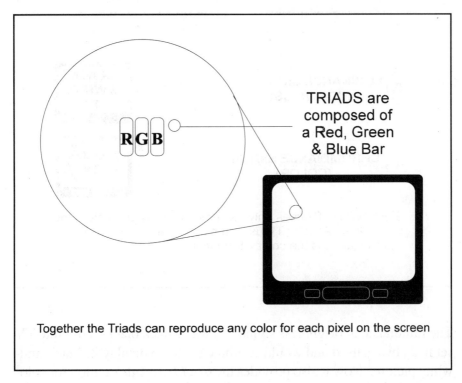

Figure 2-2. The make-up of a color TV picture.

This is possible because of the *additive color system* or *process*. It lets you mix (add) the three primary colors, red/green/blue, together at different intensities to achieve virtually any color. By varying the brightness of each red, green and blue bar, the television screen can depict any shade of almost every color known. This is called an RGB picture tube.

Example: Three bars of equal parts red, green and blue make white. If all three were less bright, a gray color would result, all the way down to black.

Luminance and Chrominance Signals. Refer to Figure 2-3. Way back when color TV was first introduced, there had to be a way to phase in this new type of signal. The solution was to allow the older black-and-white TV sets to continue to receive the information they needed to recreate the images. So broadcasters transmitted a signal that was broken up into two parts: *brightness* (luminance) and *color* (chrominance).

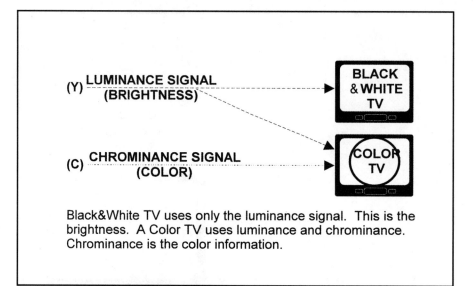

Figure 2-3. Luminance and chrominance: color TV is born.

The luminance or brightness signal (Y) was what a black-and-white TV set had always used and would continue to use to display its black-and-white picture. Now it also provides the *brightness* information for color TV sets.

The chrominance (C) signal contains *hue* (the actual color, such as red) and *saturation* (the depth of the color). With this example, saturation would make the color either a light or deep red. The chrominance signal is ignored by a black-and-white set. However, the Y and C signal together form the signal that feeds your eyes the true tints, tones and textures of a color TV.

HOW IMAGES ARE CREATED ON THE SCREEN

The television tube is actually an optical illusion, a trick to the human senses. The picture on the screen is "painted" with *ONE* fast-moving point of light. See Figure 2-4. An electron gun shoots electrons on a phosphor-coated glass screen. The phosphor glows when struck with the single point of electrons. Note that black-and-white TV requires only one electron gun. Color is a combination of three guns, or one gun with three beams.

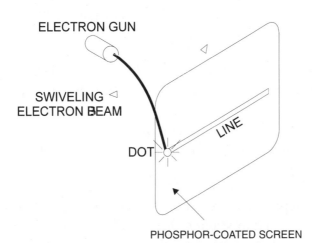

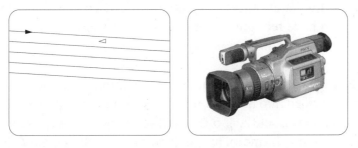

A DOT is created from the electron beam hitting the phosphor-coated screen. That dot forms a line when moved back and forth at great speed.

The lines are drawn from left to right, then the beam comes back and starts the new line. Together, the lines form a picture.

Figure 2-4. How images are created on the TV screen. Drawing by John Adams. Camcorder image reproduced with the permission of Sony of Canada Ltd.

Imagine taking a sparkler in hand and waving it back and forth at great speed. What do you see? If you are moving it fast enough, it will form a *phantom line of light*. ONE single point of light will create an illuminated line. This is simulating an electron gun's actions.

Imagine this: if you were the electron gun, at the end of each sparkler wave, you drop down a fraction of an inch and begin another line. One line after the other. After you get too far down, you start at the top again and wave your arms. Over and over. If you are able to do this at a tremendous speed, you would have a screen of pure light in front of you: a wall of light created by one glowing dot.

By stringing the images together, one after the other with a slight change in each image, you would create the illusion of motion. Television gives this illusion by painting one picture, then another and another. Each image has a slight difference, thus motion is created. Just like an animated cartoon. On a U.S. TV's tube, a new picture is painted every thirtieth of a second.

NOTE: Computer monitors don't use the interlacing scheme because humans can actually perceive flicker at anything less than 50 frames per second, especially if something is moving across the screen at the time. It is possible that the new DTV format will make use of non-interlace schemes (60 frames per second) for our new TVs. This means we will get a complete picture for each frame. The results are stunning!

The sparkler example relates to your TV like this: The point of light would be the electron gun striking the phosphor-coated screen. The arms that wave it back and forth, up and down, would be the *yoke*. This is done magnetically.

One difference between color and black-and-white television is that the intensity of the beam of light can be varied electronically, darker at times and lighter at others. The other difference is that a color television would have red, green and blue sparklers taped together to create color, using the additive color system.

Interlacing, Fields and Frames. Each frame of a picture on a TV screen is divided into horizontal lines. There is an odd set of lines (1, 3, 5, ...) and an even set (2, 4, 6, ...). All of the odd lines are scanned onto the screen in one frame. This "half frame" is called a *field*. The electron gun comes back up to the top and paints all of the even lines. This is called *interlacing*. Each field takes one sixtieth of a second to scan onto the monitor. Both fields together (odd and even) take one thirtieth of a second, which is called one *frame*. On an NTSC TV, there are approximately 250 scan lines per field, two fields per frame and 30 frames a second.

The whole scheme of interlacing is possible because the human eye can retain an image for 80 milliseconds. So, while the odd lines are being painted, the last set of even lines is still on your mind, so to speak.

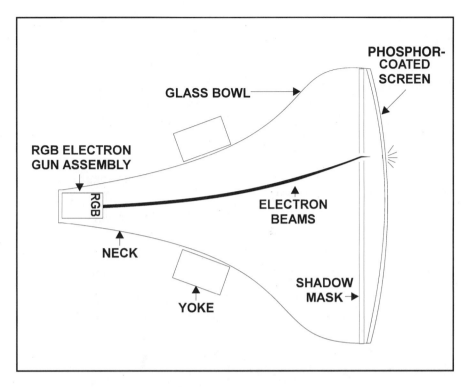

Figure 2-5. Components of a modern picture tube.

COMPONENTS OF A MODERN PICTURE TUBE

By looking closely at the components of a modern tube, you will be able to see just how it creates magic. Refer to Figure 2-5.

The tube itself is an air-evacuated glass bowl and neck. It plays host to a stunning array of components. Starting from the back of the tube (the neck), we have an electron gun assembly. A color television has three guns, or one gun with three beams. Each emits a high velocity stream of electrons which are accelerated through the neck with a combination of various electrical phenomena. The electron beams are then focused and directed with electrodes and electrostatic fields. As they come through the tube, a powerful set of electromagnets called the *yoke* swivels the beams horizontally and vertically to create the scan lines. It's quite a hummingbird-like task, especially at the speeds needed to make the beam fly back and forth.

Before the electrons hit their target, they are highly localized with a grate-like device called a *shadow mask*. It is a stencil, of sorts, that lets

NOTE: Most screens are composed of dots in the shape of a square. The squares are laid out like a brick wall turned on its side, or in some cases exactly like a screen from a window.

the beam spray only the intended targets ahead on the glass faceplate. See Figure 2-6.

The faceplate, or picture screen as we know it, has hundreds of thousands of dots painted on it. Each dot is composed of the three phosphor-coated bars described earlier. There is red-producing phosphor, as well as bars of green and blue.

When the corresponding color's electron beam finally makes it to these phosphorous bars of color, the image appears on the other side of the faceplate, giving us a spectacular view of the video universe.

As you can see, the picture tube is not nearly as complicated as you may have expected. This is the same way black-and-white TV was produced 40 years ago, except it was with one electron beam that controlled only luminance. Today's TV sets have improved upon each element of the tube.

IMPROVED TV TUBE TECHNOLOGY

Picture Quality. When customers started to demand a crisper, brighter picture, companies were forced to deliver. The old WWII complacent

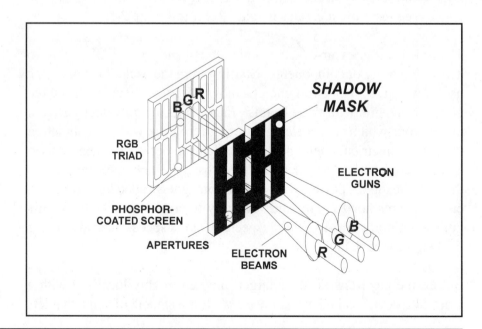

Figure 2-6. How the TV screen's shadow mask works.

Howard W. Sams & Company **Complete Guide to Video**

technology was no longer enough for viewers. As a result, new materials and sciences have jumped TV tube technology into light speed. Darker picture tubes produce a more contrasting picture in daylight. A flat-front screen all but eliminates glare: great for bright rooms. Invar shadow masks allow you to run a tube hotter without any focus or age problems. (This means a brighter, more vivid picture). Rare-earth elements are used in the phosphor which let the picture tube display ever brighter images for years and years without fading. In the immediate future, the biggest improvement will be HDTV signals. They will produce the most lifelike images ever seen through a bowl of glass.

ALTERNATE TV PICTURE TECHNOLOGY

This is the stuff video dreams are made of! Who would honestly turn down a bigger television screen? If your obsession for a larger diagonal screen measurement overcomes your sanity, then a bigscreen, projection TV or, yes, even an LCD hang-on-the-wall unit, will be the only cure. Perhaps your insanity has led you to ever smaller sets, to the point you are carrying a pocket LCD TV around, repeating to yourself, "*Jeopardy* is on in 5 minutes!" Either way, engineers have the right dose of sanity (or insanity, whichever the case may be) for your video ailments.

Projection TVs. With the massive diagonal measurements that consumers are demanding, a larger, heavy glass picture tube is no longer possible. The solution for 40 inch-plus TV sets is to use three high-power picture tubes (red, green and blue) with built-in phosphor screens to *project* the image onto a neutral screen. See Figure 2-7. These are called *Schmidt projection tubes* with an *internal screen*. The three units run at sun-like temperatures because of the high brightness needed to project the images over a distance. Once the beams are put through the phosphor coating, they are focused and corrected with a special lens system called the *Schmidt optical system*.

Rear Projection TVs (Bigscreen TVs): These are one-piece units with all parts in one cabinet. Three high-output tubes send their three pictures up to a mirror then forward them through a nearly transparent screen. There is a red picture, green picture and blue picture. Simple.

NOTE: Degaussing will cure color purity problems. Color problems are indicated when colored patches appear on the screen, or discolorations are visible on a white background. Whenever you move a TV, it should be degaussed as it needs to readjust to the Earth's magnetic fields. This is easily done by using the TV's internal degaussing circuits. Turn the TV on for one minute and off for 30 minutes. Repeat until the color purity is strongest.

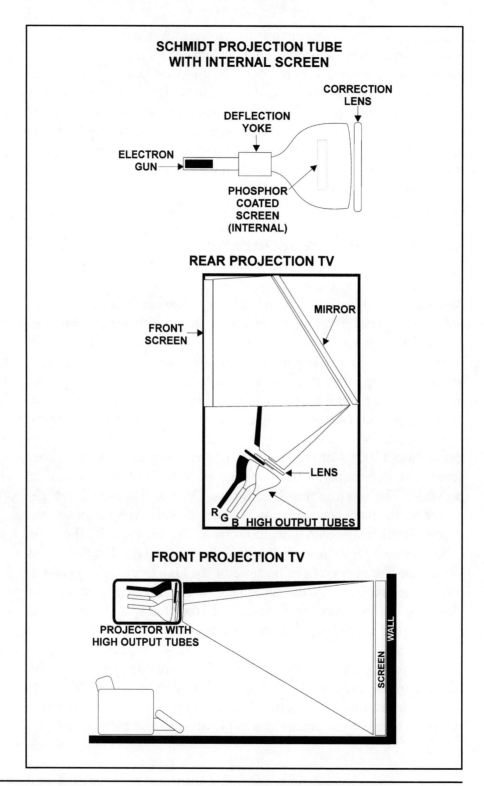

Figure 2-7. Projection television technology.

Front Projection TVs: These units are the same as the rear projection TV, except they are in two pieces. The high-output tubes are placed several feet from the screen itself. The major limitation for front projection TVs is that you need almost total darkness to watch the afternoon ball game.

LCDs and Gas-Plasma Screens. (Also known as *matrix displays*.) These units have hundreds of thousands (and in some case millions) of individually controlled pixels. The computer in the TV will address each pixel and make it the correct color.

LCDs: These are called non-emissive displays as the pixels do not emit light. They simply act as a sort of *light valve*, letting light through from the back. LCDs range in size from 1" to a unit which is now pushing 42" but costs $25,000. Sizes up to 60" for under $2,500 will soon be possible.

Gas-Plasma: These are called emissive displays as they emit light for each pixel. This is quite a new technology, but keep an eye on it. A 40" set is currently around $15,000! This is due to the manufacturing process and the amount of throwaways just to get one working set. Gas-plasma and LCD are likely to replace the bulky TVs of today with units that are merely inches thick and can hang on a wall (just like in the movie *Total Recall*!). I recently saw what a few manufacturers had to

---

TV SCREEN FORMATS

TVs, camcorders and other video equipment can display or record many different picture formats. The standard TV screen format is 4:3. This refers to the ratio of the screen's width to its height. If a 4:3 screen is 20" wide, then it is 15" high. This is a 25" diagonal TV screen. Another ratio used in HDTV and other video elements is 16:9; 16 units wide and 9 high. This is also called *widescreen*. This format corresponds more naturally to a human being's visual field. In fact, most movies are shot in some form of widescreen format. When a 16:9 movie is broadcast onto a normal TV (4:3), it has to be converted in one of two ways. *Letterbox* shrinks the 16:9 picture and places a black space above and below. Annoying! *Pan and scan* places a theoretical 4:3 screen over sections of the picture and cuts everything else. Approximately 25% of the widescreen picture information is lost. A 16:9 TV screen does not need a conversion, and can take a letterbox movie and expand it to full screen.

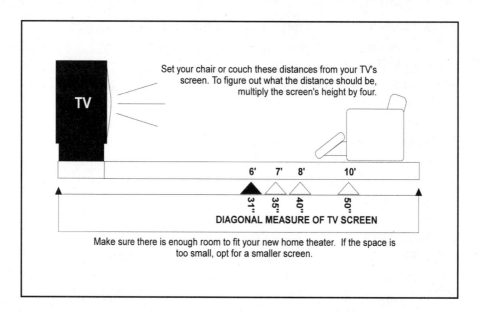

Figure 2-8. Television placement in a room.

offer in this field, and was simply stunned at the quality. For $15,000, I should hope so!

CONSIDERATIONS WHEN CHOOSING A PICTURE TUBE

Now that you know the improvements made in picture tube technology, let's look at the TV purchasing considerations you need to know in regard to the picture tube.

Where the TV will be placed will determine the tube's size. See Figure 2-8. If you only have a 9' long room and will be sitting 6' from the TV, then a monstrous 60" projection TV is complete overkill. As you know, TVs are generally classified by the diagonal measurement of the picture tube. The most obvious improvement of televisions are with respect to the tube's size. Most companies produce TV sets in 13", 19", 25", 27", 31", 32", and now up to 40". Bigscreen and projection units top out at 120" or so, and are ever increasing in size.

A few basic considerations are: "Will this TV fit in my entertainment cabinet?" "Is the set too heavy for me to move around?" "Does its exterior color and styling fit the decor?" Thirty-two inches of pure TV technology is great, but it can weigh up to 300 pounds. Heavy prices

DTV and HDTV

Analog picture quality can only advance so far before digital technology is needed to replace it. The amount of information needed to display an NTSC picture is 5 to 10 times that needed by a digitally-compressed picture. The antiquated analog broadcast standard (NTSC) is no longer viable in today's age of microprocessors and digital technology. So, in 1987, the FCC set out to define a new digital standard. It would be equivalent to updating AM to FM or vinyl records to CDs.

High definition television (HDTV) and digital television (DTV) are *NOT* synonymous. HDTV is a quality, high-resolution (1000+ lines) TV standard that offers 4:3 and 16:9 formats. It implies a much sharper-quality, wide-screen picture. However, *DTV* only implies that the picture is in a digital format and not necessarily high-resolution or wide-screen.

Digital TV takes advantage of the fact that a digitized picture can be compressed into a much smaller space. By using compression, the digital picture takes much less bandwidth to transmit. This allows you to pack in six times the information of a regular TV's picture and sound. That would be enough to send an HDTV signal that is up to 1300% sharper, OR it would support six times the amount of channels in the same bandwidth currently used to send one analog signal.

What does all of this mean to you, the consumer? It means that you will soon need either a DTV television ($2000±) or a set-top converter ($200±) to watch TV. The new DTV format will be phased in and will eventually replace the NTSC broadcast standard. On April 3, 1997, digital television standards were finally adopted by the FCC. The top ten broadcast markets are supposed to be transmitting this signal (as a simulcast on a separate channel) within two years, although they have pledged to do it in 18 months. The other top 30 markets are to follow in just under three years. In the year 2006, all NTSC signals are scheduled to cease. By then all broadcasts will be in this new format. This means you will either need a special decoder hooked to your current TV, or a DTV set. Standards are not completely set, so it is not known what the final costs will be.

THE HDTV/DTV CONTROVERSY
There is a serious controversy over the new DTV standard. Here it is in a nutshell: In 1987, the FCC wanted to set HDTV standards to compete with Japan. The Japanese have had HDTV for years. The original concept was to have broadcasters send higher-quality digital pictures to our homes. However, the original HDTV idea has now been heavily twisted.

The broadcasters argue that people do not want a better quality image, so why make each TV station upgrade their video cameras, tape players, etc.? What the broadcasters have neglected to take into consideration is *people DO what a better picture*; much the same way we prefer stereo over mono.

DTV was the compromising answer to HDTV. One low-quality NTSC analog signal takes up 6 MHz of bandwidth to transmit. So does one *high-resolution* HDTV signal. The FCC decided to *give away* 6 MHz of bandwidth (a UHF channel) to each broadcaster, to simulcast programming on an NTSC signal AND a DTV signal.

However, the broadcasters had another idea. They want to use the new bandwidth to transmit one low-resolution digital TV channel (NTSC quality) with one of the 6 MHz, and use the remaining 5 MHz of space for other services: pager services, Internet access, software transfers, stock quotes, or other profitable ventures that use radio waves. In other words, the FCC gave broadcasters 11 to 70 billion dollars-worth of TV channel bandwidth for their own profiteering schemes. What is the outcome of this? We are not likely to see improved quality HDTV, and to make matters worse, we will have to upgrade our own video equipment for nothing! No better-quality pictures.

NOTE: As of writing, the FCC has left outlining the standards to the broadcasters themselves. Unfortunately, they cannot agree on what the new DTV format will look like. Ten years after this TV technology race started, we are still awaiting the final format.

typically go with the heavy equipment. Also, you should consider this: "Can I justify this expenditure, or should I come back to the reality of a mortgage and kids?" Doing away with some features will lighten the last consideration. Also, consider the illumination of the room. Is it a sunny room or a dim cave? Keep all these things in the back of your mind while purchasing your new TV set.

Now think about the physical placement and use of the TV.

Living Room, Family Room or Home Entertainment Center. Obviously, this is where the largest TV set in your house will reside. You have two choices for this area:

Tabletop Model: The average living room sports a 27" modern TV; plenty of screen. Even the most discerning videophile can have their visual palette satisfied with this size TV, without a price tag that would choke them to death. At the time of this writing, a typical 27" TV with a modest amount of features goes for about $450 - $600 retail. If you would like more screen and are willing to pay, go for it! I don't know anyone who wouldn't take a 32" TV over a 27" if they could. Just don't get rid of the house to pay for the equipment.

Bigscreen or Projection TVs: If you have a dark living room or TV room, have the space and really want to immerse yourself in the picture, go for a bigscreen TV. A 50-52" model will work fine. Expect to pay $1500 - $2500 for it.

Kitchen. A 13" to 14" TV set is fine for a bit of entertainment while cooking. It's better than watching the microwave spinning leftovers. If you spend a lot of time in the kitchen, a TV/VCR combo might work for you. You can watch a talk show and record a soap opera at the same time. This part of the home, or possibly a camper, is the only advantageous place to have one of these combo units; otherwise try to sway your urge to purchase one. It is just as cheap, if not cheaper, to buy the VCR and TV separately. Expect to pay $125 - $200 for a 13" to 14" TV set and around $300 - $400 for a TV/VCR combo.

Kids' Room or Recreation Room. A 19" TV is fine for these areas. Look for A/V jacks on the front panel, as the Nintendo is certain to be a permanent peripheral attached to the TV set. Face it, moms and dads, the game console is here to stay. Get the kids their own TV and keep the home entertainment center free of channel battles. Nineteen inches is the sweet spot for TV tube sizes. There are plenty of mass-market 19" TV sets going for $175 - $250.

Master Bedroom. A 19" - 25" will provide great entertainment by which to fall asleep. You want a system that doesn't take over the room. Usually, people just want to relax with the news or the late-night movie before going to sleep. A 19" TV is usually plenty of screen for the average master bedroom's size. Don't go overboard while looking for a bedroom TV. A cheap mass-market TV set will suffice.

FEATURES EXPLAINED

Resolution. The general purpose definition of *resolution* is: "The measure of a picture tube's ability to reproduce fine details. Simply stated,

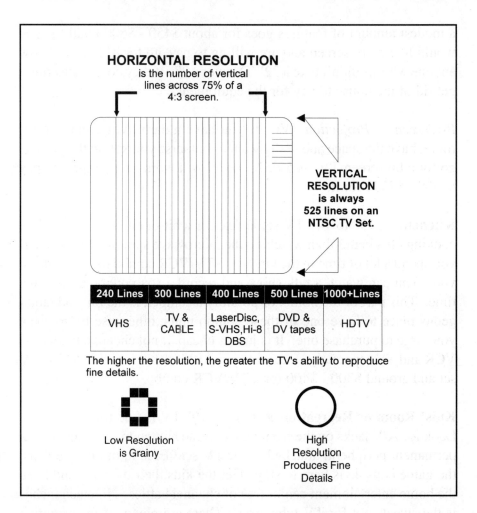

Figure 2-9. Resolution of a television screen.

it's clarity measured in scan lines." The *grainy look* to a picture on a TV is due to low resolution. A high-resolution TV is capable of a near-photographic image.

Horizontal Resolution: The number of vertical lines that can be displayed in a picture width equal to the picture's height (counted on a horizontal axis). The greater the number of these lines, the sharper the picture. See Figure 2-9 for a pictorial explanation. Cable and VCRs will provide a signal to display under 300 horizontal scan lines. Hi-8, S-VHS, DBS and laser disks can display 400 lines. The new digital video disk format provides a signal that can hit 500 and up. For the

> **ADJUSTING YOUR TV'S CONTROLS**
>
> Finding your way through a maze of on-screen menus or dials for adjusting your TV's picture can be a chore. However, it is well worth learning how to set a perfect image that is pleasing to your eyes. The company's factory settings are not set in stone. In fact, everyone has a different preference as to contrast, brightness, hues, etc. It's not a hard task to figure out your TV's controls; all you need to do is use this sequence:
> 1. Adjust the *brightness* (black level) and *contrast* (picture). These are the luminance controls.
>   - Display a black screen.
>   - Turn the *contrast* dial or bar to the minimum setting.
>   - Adjust the *brightness* dial or bar to reproduce the black correctly, just slightly above pure black and not a shade of gray.
>   - Display a picture with white in it.
>   - Turn the *contrast* up until the display has the brightest white you desire.
> 2. Adjust the *color*, or chrominance next.
>   - Display a set of color bars such as the ones used by local stations when they go off for the night. Adjust the color until no there is no "blooming." Blooming occurs when the red and magenta bleed into each other (2nd and 3rd bar from the right).
> 3. Adjust the *tint* (hue).
>   - Do this by looking at the yellow bar, second bar from the left. Make sure it is a true lemon yellow color and not lime, green or orange.
> 4. Turn on a station and see if skin tones are being displayed properly. If not, repeat the above steps.

ultimate, one of the HDTV formats proposed will be 1,000 and up. This is pristine clarity and sharpness at its best.

Vertical Resolution: Because an NTSC signal is made up of 525 horizontal scan lines per frame and 30 frames per second (250 lines for the odd lines, 250 for the even lines and 25 for the closed captioning, color info and a few other bits of information), it is only capable of a vertical resolution of 250 scan lines per field. A PAL signal is capable of 300 scan lines per field. Until the NTSC standard is changed to HDTV or DTV, this figure will remain the same for every TV set.

Purchase Advice: Horizontal resolution is the number you need to examine when purchasing a TV. Look for a set over the 500-line range as it will be able to take full benefit of new DVD technology. If you plan on upgrading to HDTV or DTV in the future, and don't mind paying the big-ticket price, look for sets above 500 lines. Don't hold your breath for the HDTV format, though. (See *DTV AND HDTV* for more information on this.) If you own a bigscreen TV, a low-resolution signal such as from a VCR is quite visible. The higher the resolution, the more details you will see in a picture; so try to feed your TV high-resolution signals. See more about resolution in Chapter 3.

NOTE: The controls on your TV will include either *Contrast* or *Picture* functions. They are synonymous.

Contrast. Contrast means the difference between the brightest and darkest areas of a picture on the screen. Images are more vivid, lively and pleasing to the eye if they contain high contrast; which is why this book is printed with black text on white paper. Modern televisions use electronic and mechanical improvements to increase the contrast and thus the ease of viewing. Manufacturers now tint the TV screen's glass so a darker black is possible. This contrasts the whites even more. Light sensors also adjust the contrast levels to compensate for the room's light.

Brightness. This is a function of your TV set's contrast. Another word for brightness is *black level*. Shadow mask technology is allowing whiter whites to be reproduced and displayed. See *Adjusting Your TV's Controls* for an explanation of how to adjust your TV's contrast and brightness.

Focus. A TV set focuses its image in two ways: mechanically with a shadow mask, and electronically with magnetic, electrostatic and electrical fields.

Shadow Mask. Refer back to Figure 2-6. This part of the tube has had a recent infusion of high-tech medicine delivered to it. The job of the mask is to aim each color beam directly onto its intended target, the phosphor. The mask also has the ability to narrow a wide electron beam. Look for a TV set with an *invar* mask, as it allows a hotter beam to be run: meaning a brighter picture. Some sets also have a special black coating around each phosphor element. This reduces the amount of reflected light and thus increases contrast.

Comb Filter Circuitry. This will increase resolution and eliminate extraneous color in the picture's detail: in other words, it increases sharpness. The comb filter does this by providing better separation of the luminance and the chrominance signals. Most 27" or larger TVs have this feature. If the TV you're examining does not, demand a TV that does.

Flat Screens. For anyone with a flat-screen picture tube, there is no going back to a regular curved screen. Flat screens all but eliminate glare and reflections. They also reduce distortion at the picture's edges.

For more features and purchasing hints, see chapters 8 and 9.

SOUND

If you are a viewer who covets sound, can you image watching *Star Wars* without hearing a single laser-cannon blast? How about *Casablanca* with the absence of even one note of music? It would be sacrilegious to say the least. The sound circuits in a modern TV deliver half the experience of a movie director's creation.

How do the sound circuits work? How can you drive speakers to rock the house with thunderous bass explosions? How do you make a sound field to immerse your mind in memorable movie scenes, even if only to hear a romantic secret whisper? What are Dolby Surround sound and THX anyway? Is there any new technology on the horizon? On we go with *sound* explanations.

THE SOUND EXPERIENCE

Most mid- to upper-range television sets come with stereo sound and built-in speakers. Realize that even the best of these TV sets doesn't compare with a cheap component stereo. This is fine for a small TV set placed in a bedroom or kitchen. However, your home theater can be vastly improved with a separate stereo source, and a fair-sized amplifier and speakers. This will give you a *sound experience* as opposed to straining your ears on a small TV's inadequate speakers.

SOUND OPTIONS

Your sound options expand as the technology grows. Currently, 13" - 20" sets are typically equipped with one-dimensional, ho-hum mono sound and a few built-in ultra-cheap speakers. Some 19" sets do come with an audio jack into which you can hook your stereo.

MTS/SAP. Your next option, on a slightly higher scale, would be a TV set with a built-in multichannel television sound decoder (MTS) and a few 2- to 3-way speakers. The separate audio program (SAP) feature lets the broadcaster send an alternate language soundtrack to your set. These features are typically on mid-range 27" sets, making it great for an average home theater set.

Surround Sound. Some higher-end TV sets have a simple built-in surround sound feature. It lets you hook in a set of surround speakers. The two other channels are voiced through the TV's built-in speakers; though some units allow externals as well.

SOUND TERMS

Let's take a deeper look at a few sound terms:

Stereo. The derivation of *stereo* is "three dimensional." See Figure 2-10. Stereo is simply a way to reproduce sound in a way that is conducive to the biomechanics of your ears. It uses *multiple* recording sources (microphones), then echoes these same recorded sounds through *multiple* speakers. In this way you get a simulated 3D sound field; much

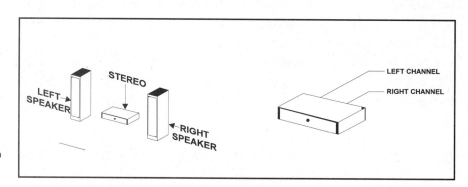

Figure 2-10. A standard stereo setup.

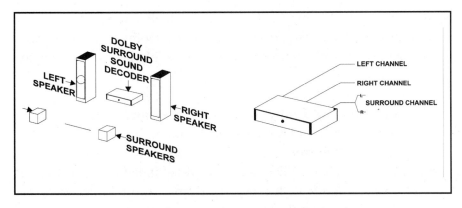

Figure 2-11. A surround sound system.

the same way a 3D visual image is created on a two-dimensional surface. Without this method, you get only a mono effect. Boring!

Surround Sound. See Figure 2-11. Stereo uses two sound channels, right and left. When we refer to *home theater stereo*, the industry has come to preach a different definition: that of *surround sound*. This is a way of "surrounding" a person in a totally directional sound field. It usually take four or more speakers to emulate a real *sound stage*.

Close your eyes. Imagine you are sitting still in the center of a room. A person is walking around you, talking. Even without looking you can determine where he is in the room. This is because a human ear is capable of determining the direction from which a sound is coming. Surround sound systems can reproduce this sound scenario. More on surround sound later.

MTS/SAP. Multichannel Television Sound. Second Audio Program. MTS is usually found on mid- to high-range television sets. It utilizes a decoder that lets you receive stereo sound, then amplifies it to your TV speakers. Most of these sets also allow you to hook up the MTS to your external stereo. MTS is a way to save on having to purchase extra stereo equipment for your TV: great for a 19" - 27" TV. The other advantage is that you can receive a foreign language audio track that is simultaneously broadcast with the English language track.

Dolby. A technology licensed by Dolby Laboratories, Inc. Dolby's technology reduces *system noise* in a sound system by using compres-

NOTE: Dolby's Website has great information on setting up home theater sound systems and other home theater information: http://www.dolby.com/

sion/expansion technology. Their multichannel stereo sound technology is also what makes surround sound and Dolby Digital possible. Dolby does not manufacture equipment; it merely licenses its technology.

Dolby Surround and Dolby Surround Pro Logic. This technology encodes four channels of sound while a movie is being filmed; left, right, center and a surround channel. To listen to the signal in your home theater, purchase a Dolby Surround-equipped decoder/receiver, A/V receiver, or a TV with a built-in Dolby Surround decoder. See Photo 2-2. This decodes the Dolby Surround signal and sends it to your speakers to create a dimensional sound effect.

A basic *Dolby Surround* decoder takes the original encoded signal and changes it into three channels. One is used for the left stereo channel, and another for the right stereo channel. The last channel is used for surround (the two rear speakers). With Dolby Surround, the two front channels combine to output the center channel as a "phantom" dialogue channel. The problem is that the phantom center channel can only be heard if you are centered between the two front stereo speakers. The surround channel is for effects and atmospheric sounds.

A *Dolby Surround Pro Logic* decoder changes the signal into the original four channels. See Figure 2-12. Two speakers are for the front left and right stereo channels. Another is the surround channel. Now, instead of a phantom dialogue channel, the Pro Logic decodes an actual fourth center channel and sends it to a speaker, usually situated above

Photo 2-2. A hi-fi receiver with a Dolby Pro Logic decoder. Reproduced with the permission of Sony of Canada Ltd.

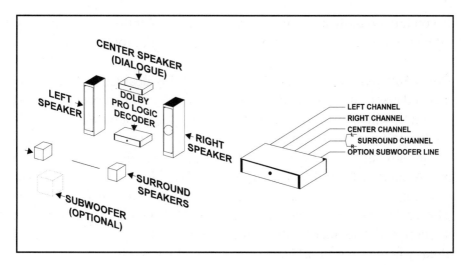

Figure 2-12. Dolby Pro Logic surround sound.

or below your television. This channel gives a feeling that the speaking character or the sound effects are directly in front of you. The advantage of the Pro Logic is that you can be off to the side and still receive the dimensional effects of the center channel. Remember, Dolby Surround's phantom center channel lacks directional capability. The Pro Logic's separate center channel and speaker are well worth the extra $75 or so. You may consider buying the decoder now and adding the speaker later. Dolby Pro Logic decoders allow you to switch off the surround channel and use only the TV speakers until you are able to add rear speakers. It is called the Dolby 3 Stereo mode.

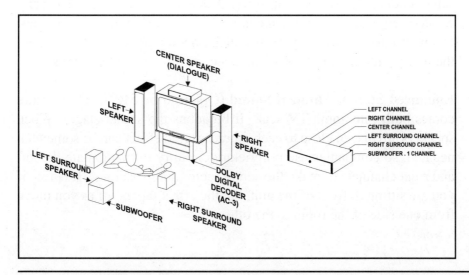

Figure 2-13. Dolby Digital surround sound.

Chapter 2: Sights, Sounds & Signals: All About TVs

If you are a bass-head, you may want to add a satellite low-frequency effect subwoofer to the Pro Logic system. Time to rock the house.

Dolby Surround AC-3 or Dolby Digital: Dolby decided to go one further and made an encoding/decoding system with 5.1 independent sound channels, which bring you an ultra-realistic sound stage. See Figure 2-13. There are two front stereo speakers, and the two rear surround speakers are now in stereo where the single mono-surround channel used to be. A center (dialogue) channel is standard. Now there is a separate low-frequency bass effects channel that adds acoustic *punch* with a subwoofer; this is the .1 channel. The new DTV broadcast standard and the new DVD standard will push Dolby Digital into full swing. It is already being used in movie theaters around the world.

The whole idea behind Dolby Surround, Dolby Pro Logic Surround and Dolby Digital is to bring you the sensation that you are in a movie theater immersed in the sound effects. Quite an achievement to cram thousands of square feet of sound space into a living room. However, will it ever replace the movie theater experience?

THX. THX processing emulates the effect of a large theater without actually simulating its acoustic reflections. It is not a replacement for surround sound: merely an enhancement. THX is a licensed technology that sets out hardware standards for home theater equipment. A THX package comes together to form a system designed to reach the same loudness levels used in movie theaters. George Lucas basically wanted people to experience the movie theater feel in their own homes. Is it worth the money? If you are a diehard videophile, sure! If not, put the money into quality Dolby Pro Logic equipment and speakers.

Enhanced Stereo, Ambient Sound (Q-sound or SRS). This feature comes built into some TV sets. It broadens the sound stage. When speakers are placed close to each other, a stereo effect can be somewhat negated; so companies use proprietary circuitry to manipulate the left and right channels to make the sound seem more spacious. It is fine if you are sitting in front of the unit, but the effect tapers off as you move from one side of the room to the other.

WHAT CONSTITUTES AN AVERAGE HOME THEATER SOUND SYSTEM?

What system should you buy? Ask yourself if you want a total teeth-shattering sound system or simply a way to make out an anchorperson's voice. Do you want it to sound as if you have great seats at a symphony? Maybe action movies with their surround sound tricks excite your ears. Look at the definitions and explanations of your options and decide.

If cost is stopping you from having the home theater sound system of your movie dreams, then buy one component at a time: a receiver here, a set of speakers there, etc. Purchase a surround sound TV that has A/V outputs, and plan for the future. Even if you don't have an amplifier and speakers, at least you are well on your way to having an amazing home theater sound system. Most 27" TV sets come with a minimum of MTS/SAP or enhanced stereo. Try to steer away from this as most of the new technology coming out involves digital surround sound. Spend a little extra now and save yourself from having to buy a whole new set in two years.

If you have decided on surround sound, buy a minimum of a Dolby Pro Logic A/V receiver. Try to get a Dolby Digital (AC-3) unit if possible, though.

EQUIPMENT

If the features you want are not built into your TV set, you may opt for a complete home theater sound system. They can be purchased as separate components or as bargain packages. Here is a list of components:

Decoder. This component decodes the Dolby Surround, Dolby Pro Logic or the Dolby Digital signals.

Receiver. Just like its name, the receiver receives the audio or video signal.

Photo 2-3. Home theater central surround speaker package. Reproduced with the permission of Sony of Canada Ltd.

Power Amplifier. The *amp* is used to amplify (add power to) the output signal going to your speakers.

Speakers. These act as a transducer that turns the electronic signals into sound. See Photo 2-3.

Some units have the decoder, receiver and power amp built in. Others have additional features such as remote control, THX, etc. Be aware of what you are buying because there are SRS (enhanced stereo) units that are NOT surround sound but may appear to be. See Chapter 8 for purchasing strategies and Chapter 9 for further features.

FINAL SOUND BITES

Gone are the days of mono speakered TVs. Customers now realize that sound is half the experience. Why should anyone have to go to a movie theater just to get that awesome sound encounter? Now it is affordable for our homes. *Sounds* good to me!

See *Howard W. Sams & Company Complete Guide to Audio* for more home theater sound information.

SIGNAL

If the TV's picture tube and sound system make up the heart of video, then the *TV signal* is the soul. Together they make movie magic happen. They both have the task of entertaining and educating your senses and feeding your personal tastes. What exactly is a TV signal and how does it bring you your favorite shows? How does the TV receive the signal?

Quite simply, a TV signal is pure information. It contains the picture information, timing information, color information and audio information. TVs are useless without this signal (information). The quality of the signal makes or breaks a fantastic video system. This section will touch briefly on this subject, but see Chapter 3 for the practical side of signals (antennas, cable boxes, satellite receivers).

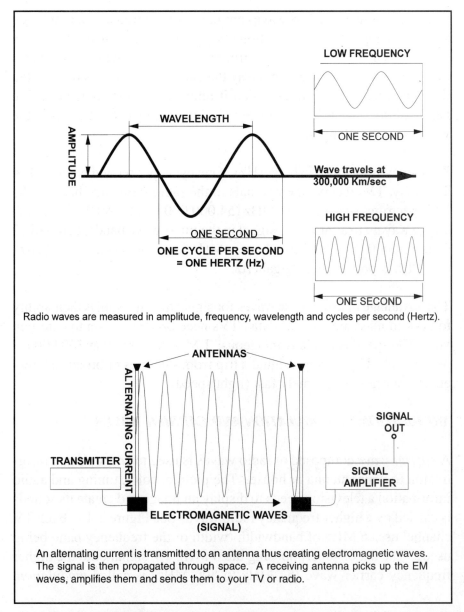

Figure 2-14. How electromagnetic waves are measured.

EM WAVES

A TV signal (information) is sent in the form of *electromagnetic waves* (EM): radio waves, to be exact. EM waves are basically energy delivered from one point to another at the speed of light. They can be anything from TV signals (radio waves) to light, to gamma rays. EM waves are used to *carry* sight and sound to your TV.

Refer to Figure 2-14. EM waves are measured in frequency, amplitude and wavelength. Each alternation (cycle) of current flowing through a wire emits a single wave of electromagnetic energy which propagates through the air; much the same way the energy of a wave is propagated through water. The rate at which it alternates is the frequency. The height of the wave is its amplitude. The length between the crests of the wave is its wavelength.

Frequency is measure in hertz (Hz). One cycle per second equals 1 Hz. Typically, TVs receive their signals in the radio frequency band of the EM spectrum, starting at 54 MHz (54,000,000 Hz) for VHF. This goes all the way up to a satellite's signal in the microwave band, at around 12 GHz (12,000,000,000 Hz). Note: MHz is pronounced "megahertz" and GHz is pronounced "gigahertz."

The advantage of using these waves for signal transmission is their ability to pack in mass amounts of data. TVs need *oodles* of data to entertain you. The speed is also advantageous: EM waves travel at 300,000 *km* per second. This makes a signal's trip from satellites or broadcast towers to your home extremely fast (light speed).

MODULATION/DEMODULATION AND CARRIER WAVES

A *carrier wave* composed of radio waves is used to *carry* picture information to your antenna or house. The picture, color, timing and audio information a television needs to display an image and create its sounds is carried by a higher frequency radio wave. See Figure 2-15. Each TV channel uses 6 MHz of bandwidth (width of the frequency band being used). The act of superimposing the 6 MHz TV signal onto the higher frequency carrier wave (54 MHz to 806 MHz) is called *modulation*.

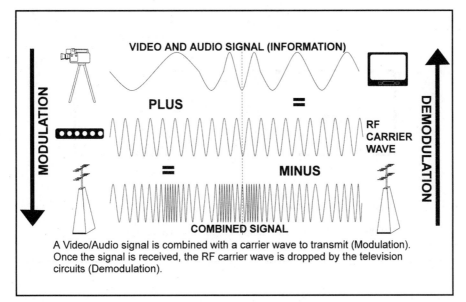

Figure 2-15. Modulation and demodulation of a television signal.

Sort of like a piggyback ride. This creates the *broadcast signal*. After the signal reaches your TV's tuner, the carrier frequency is discarded, leaving the original information for use by the TV's circuits. This is called *demodulation*.

WHAT MAKES UP A TV SIGNAL

Refer to Figure 2-16. There are basically two video signals of interest. One is the broadcast signal. This is the modulated RF signal received from a broadcast or cable company, and it is usually carried around your video system through a coaxial cable or twin cable (often called an RF signal or RF cable). The second is the baseband video and audio signal. This is the signal left after the RF signal is demodulated and separated into an audio and video line. It is routed to your TV's drive circuits or to other equipment, such as a VCR, with A/V wiring. The baseband signals have the audio information on one line and *composite video signal* on the other. Together they reproduce the moving images on your picture tube and give you sound to set the mood.

Audio Baseband Lines. Quite simply, this is the audio information's baseband pathway to various components in your system, such as TV, VCR, camcorder, stereo or surround sound decoder. If a component

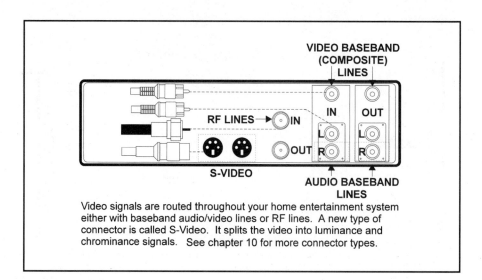

Figure 2-16. Signals in video equipment.

has one RCA audio jack, it is a mono. Two jacks mean you need to run two audio lines for stereo or surround sound.

Composite Video. This signal is made of several components: *brightness* (luminance - Y), *color* (chrominance - C) and *timing* (vertical and horizontal). Together, these form the *composite video* signal or baseband video (line-level). The composite video signal is used to route a pure video signal to equipment that uses the picture information, such as to a TV or from a camcorder.

Luminance and *Chrominance*: See the section on *Sight* in this chapter for a detailed explanation.

Timing: This signal is used to time the electron beam's movements. It is made up of horizontal blanking pulses and vertical blanking pulses. The horizontal pulses tell the TV when to sweep the electron beam across and back over the screen. This is done 15,734 times a second in an NTSC TV set. The vertical pulses sweep the beam from top to bottom then back 60 times per second (for NTSC TVs).

The tuner, IF, AGC, video and audio demodulators take in the broadcast signals and translate them into composite video and audio signals. Together they are called a *superhet receiver*.

Sound Reinforcement
Sec: 1
Sec: 15

Complete Video Guide
Chp 2
Chp 3
Chp 6

Sound Investment

Sec: 2

Sec: 15

Complete NT30 Guide

Cmp 2

1&2

3

4

Tuners. Most television sets contain a receiver unit which *locks* onto TV channels (frequency bands). The signal can then be separated into baseband video and audio. There are three kinds of tuners:

Channel Selector Dials: This is the old "click" dial. A dial is used to change the channels. They are only used on ultra-cheap sets these days, as they have inherent problems. For one thing, the mechanisms get dirty easily, making them a financial burden each time you have to get them fixed and cleaned. Second, knobs tend to fall off. This leaves you with a standby set of needlenose pliers permanently planted atop the TV set. Ugh! Third, it is next to impossible to get a remote control for selector dials. Bigger UGH!

Varactor Tuners: This kind of tuner is often seen on older VCRs and TV sets. There are 12 to 16 buttons, each with their own little dials for presetting a station. This is a very frustrating activity, as anyone that has played with varactor tuners can attest.

Phase-Locked-Loop Tuners (PLL): Also known as frequency synthesized PLL tuners. These are used on most TVs and VCRs these days. They do not need any input except for turning on the TV set. They electronically "lock" onto a station of your choice automatically. This is the autoprogram gizmo on your TV set. Just a note: For the tuner to lock onto a station, there must be a fairly strong signal present. So if you are having trouble getting your tuner to lock onto a specific channel, take a look at your antenna before sending the TV to the shop.

AGC and IF. The *automatic gain control* (AGC) adjusts the amplification gain automatically to keep a strong broadcast signal. The *intermediate frequency (IF)* is the part of the superheterodyne (superhet) receiving system that converts the higher frequencies to lower frequencies for additional amplification, filtering and eventual direction to the TV's circuits.

NTSC, National Television Standards Committee. Back in the 1950s, engineers devised the NTSC method of transmitting color television signals. It is used mostly in the U.S., Canada, Mexico and Japan. Very out of date technology-wise, but it has served us well. Refer back to Chapter 1 for a rundown on the numbers.

AIR AND CABLE

EM waves can be propagated through nearly any kind of space. Unlike sound, they do not need to sail through air; the vacuum of space does just fine. This is why satellites can send signals through space to our dishes upon the ground, and why antennas can pick up a signal without there having to be miles of wire attached to our homes. Cable, on the other hand, needs an actual connection. The waves are now electronically propagated through a metallic medium (copper, aluminum or stainless steel braided lines).

FUTURE MEDIUMS AND SIGNALS

Digital television is poised to replace the aged NTSC signals. Soon we will be thrown into the digital age whether we like it or not.

NOTE: See Chapter 10 for notes on cable and hookup procedures.

Fiber optic lines are likely to replace cable as a signal medium in the future. They will route extremely high-quality video, audio and digital information everywhere. When? It is not yet known because the cost of routing these lines to every home in America is exorbitant. You know who is likely to pay that bill.

FINAL WAVE

A TV signal is a complicated mixture of electronics circuitry and unseen waves. Just remember that it all boils down to *information*. RF broadcast signals will be addressed more in the next chapter, and line-level signals will be dealt with throughout the book. Make sure you understand both or you stand to drown in an ocean of EM wave info.

INTERFACE AND CONTROL

REMOTE CONTROLS AND MENU OPTIONS

This is couch potato heaven! Television companies have made it easier to vegetate in front of the screen with these anti-weightwatcher devices. What am I talking about? Remote controls, of course! How do they work?

Most remotes these days use an infrared *transmitter* inside the remote itself, and an infrared *receiver* inside your television or other piece of equipment. Each button on your remote beams a varying invisible signal to the TV. The TV then uses its electronic brains to decode exactly what it is you want it to do.

TYPES OF REMOTES

Dedicated Remote. This is the remote which is *dedicated* to a specific piece of electronic gear. Typically, it comes with a cheaper TV, VCR or other piece of equipment. You can use it only with the equipment with which it came. In other words, you cannot use a dedi-

Photo 2-4. An integrated remote. Reproduced with the permission of Sharp Electronics Corporation.

cated TV remote with a VCR or another brand of TV. The problem with this is that you may end up with five or six dedicated remotes just so you can watch television. Very frustrating!

Integrated Remote. This is similar to a dedicated remote, except it will work with multiple pieces of equipment; but only with the same brand name. So, if you have a Sony TV and VCR, it will control both. It saves on remote duplication. See Photo 2-4.

Universal Remote. This remote operates many types of equipment from multiple manufacturers. There are two types:

Code Entry Model: Simply enter a code for each piece of equipment and brand name in order for the remote to operate all of them. Example: Look up the codes for a Magnavox television, model TS2743C, and a Zenith VCR, model VR4106, then enter into the remote the codes that the remote manufacturer provides (note the model number of the equipment itself). Now you can control either device with the same remote.

Programmable Memory Model: These remotes *learn* from either a dedicated or integrated remote. Activate the learning mode, place each dedicated remote in front of it, then press the button on the dedicated remote along with the selected button on the universal remote you wish to program. By repeating this for each button, the universal remote will learn each command so you can use it to control your equipment.

Universal remote controls are a valuable addition to your home entertainment system. They can control the TV, VCR, cable box, and just about anything else that makes use of an infrared remote. They have typically been an after-marker item. Now, some manufacturers are offering them as an option with their equipment. Some even have a built-in VCR Plus+ programmer. Other features include:
- A remote finder. Press a button on the TV, and the remote will beep and reveal its location.
- LCD displays built into the remote.
- Lighted buttons.

See Chapter 9 for further remote control features.

Recommendations. Buy a universal remote as soon as you can. If you ever lose a specific remote, you may never be able to replace it. Also, most older model equipment is not supported by a code-entry model remote, so the old remote may be the only way to program the universal. Do it while the old remote is still capable.

ON-SCREEN PROGRAMMING AND CONTROL

The main difference between old video technology and newer technology is the on-screen programming and control functions. This displays the controls of your television onto the screen itself, either as a whole picture or simply as an overlay onto what you are watching. Thanks to microprocessor technology, you no longer have to get off the couch to change the channel or program the VCR, or see what channel is actually on. You don't even have to reach for a *TV Guide* anymore, in some cases. On-screen services offered by cable, DBS and other services will let you display on the television screen every program that is on for the next few weeks.

The whole idea behind on-screen programming is to provide an easy-to-see interface with your television. *You* are in control of the horizontal and vertical now.

Advice. Most televisions, VCRs and DBS services now offer on-screen programming or control features. When you are purchasing the equipment, try a few of the functions. Step through the sequences to adjust the picture on a particular TV and see how difficult it is to learn the functions or even how easy it is to mess up everything. Look for a set that is as graphic as possible, as it may be easier to learn. Purchase the system that makes you feel comfortable. In other words, if you want complex, then get complex. If you want simple, no-hassle controls, then get that.

WRAP UP

Television technology is forever developing. Some areas are rocketing forward while others are still moving at a snail's pace. It is up to the you, the purchaser, to demand better picture resolution, better sound and better FCC signal standards. Otherwise, it will be left to the video magnates to determine our video quality.

Now that you have an expanded understanding of the workings of your television, let's look at what you need to feed it: signals from antennas, cable and satellite dishes.

CHAPTER 3

RECEIVING WITH ANTENNAS, SATELLITE DISHES & CABLE

"One night I walked home very late and fell asleep in somebody's satellite dish. My dreams were showing up on TVs all over the world." Steven Wright

In order to watch TV broadcasts, you need a signal. Signals contain the information that converts into TV shows, movies and events. How do we get these signals?

The most fundamental (and cheapest) way to do this is with an antenna. Next is with a cable converter or satellite dish. Without one of these, you are likely to go broke renting hundreds of VCR tapes weekly just so you can watch *something* on your TV. In this chapter, we will take a detailed look at the basics of each signal receiver. As a cable customer, you may be interested to learn of new higher-quality satellite options. As a home entertainment center owner, you will want to know exactly what you need to purchase in order to receive hundreds of channels of entertainment.

ANTENNAS

Cable and satellite TV signals now offer customers hundreds of channels. Antennas are fast becoming antique curios because of this. However, do not fully discount an antenna's usefulness in your home entertainment package. Currently, you need an antenna to receive local channels if you own a small satellite system (DBS), because of legal and technical reasons. A well-tuned antenna will actually receive a better signal than your cable box for certain channels, and produce a superior visual and sound experience.

TERMS

Signal Strength. A TV antenna is rated in *signal gain*; in other words, "How far away you can be and still pick up a usable signal." In technical terms, *gain* is a measurement of the amplification abilities of the antenna, measured in decibels (dBs).

TV stations transmit radio waves, which tend to have short traveling distances or are simply shot off into space. UHF has a short range. The nature of the high-frequency signal does not allow it to bend around objects as easily as VHF. Also, the signal tends to degrade much more quickly. So the farther you are from a broadcast tower, the less power the TV signal has, requiring higher amplification (gain).

Ghosting. Ghosting is when you see two or more images on the same channel, one being transparent. Double images. Very annoying!

There are two kinds of ghosting. One is when you see a completely different channel superimposed onto the one you are watching. It is usually a grayish transparent image. This is caused when two transmitters are on the same frequency. Your signal receiver is probably in between two broadcast towers that are using the same frequency, so you simultaneously receive one strong signal and a weaker signal.

The second type of ghosting is called *reflective ghosting*. This is caused when a huge structure, such as a water tower, bounces a lagging secondary signal to your TV, which thinks it is receiving two signals: the direct wave signal and the reflective wave signal.

About the only solution to ghosting is to add a *ghost eliminator*, which is a *variable attenuator*. This device basically tones down (reduces) the signal strength and eliminates the offending signal.

Impedance. This is the measurement of resistance a cable has to an alternating current. In the case of television, it is a cable's resistance to radio waves.

This is quite important to know. The RF signal that is routed around your television is done through some form of cable. This cable has one of two impedances, measured in ohms. All flat, twin-lead cables have an impedance of 300 ohms. All TV coaxial cables have an impedance of 75 ohms. You cannot join the two without an impedance-matching transformer. See Chapter 10 for more information.

ANTENNA TYPES

Indoor Antennas. The old rabbit ears antenna and the Saturn loop of yesteryear are still floating around in the home theater market. If you are close to a broadcasting tower (under 20 miles), and there are no major obstructions in the line of sight, an indoor antenna is the best two-penny solution. Look for a unit with fine-tuning controls, built-in signal amplifiers, and a hookup for FM. A strong TV signal has minuscule degradation close to its source, so a TV-top indoor antenna will give you a decent picture. Are there broadcast towers close to you? To find out, call your local TV stations and ask about tower locations.

Outdoor Antennas. These are the insect-like metal monsters seemingly crawling atop your house. Three species are available. One is for receiving VHF (very high frequency TV stations, Channels 2 through 13). Another is for UHF (ultra high frequency TV stations, Channels 14 up). There is also an antenna that combines VHF and UHF into a neatly melded package.

Before going to Radio Shack to procure an antenna, keep in mind the range of your home in relation to various broadcast towers. Antennas are typically classified as *normal* (under 40 miles for VHF and 25 for UHF), *fringe* (40-50 miles), *deep fringe* (60-75 miles away) and *deepest fringe* (75+ miles away). Expect to pay between $75 to $150 for a VHF/UHF combo unit, or about $40 for a 120-mile UHF-only unit. If you are out in the boondocks, be prepared to buy a satellite dish.

The Design of Your Outdoor Antenna. An antenna has a mast and a boom with lateral rods pointing out. The rods vary in length as each frequency (channel) needs a different length for receiving. The greater the number of these rods, the more the signal will be received. The incoming EM waves (TV signal) are transmitted to you from a TV station's broadcast tower. They strike the rods, culminate at a central point, then are hard wired down a cable to your TV/VCR/receiver. There are four common types of outdoor antenna:

Log Periodic Antennas: If a house has an outdoor antenna upon its roof, it is likely this type of antenna. Like a millipede, the Log Periodic antenna has many legs (rods) emanating from its body. The rods are of varying lengths to pick up multiple channels. It can be used for VHF or UHF, or both depending on the manufacturer's design. You can also hook up the antenna to your stereo so you can pick up tunes on the FM dial.

Yagi Antennas: If you live in Nowheresville (50 to 75 miles from the nearest TV broadcast tower), you can use this type of antenna. All of the receiving elements on this antenna are similar in length, so you can only pick up a strong signal for one channel, a small range of channels or UHF. A *corner Yagi antenna* will allow you to receive signals from a farther distance, and are more directional.

Bow-Tie Antennas: This one looks like a stack of bow ties pinned onto a rod. It is used to receive UHF from a long distance: 40 to 70 miles. Using a rear reflector will help to strengthen the signal as well.

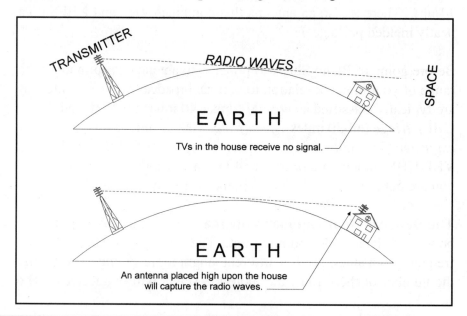

Figure 3-1. Antenna placement.

Dish Attachment Antennas: These attach right to your small satellite dish so you can receive local channels. Very handy and easier to install than a full-blown Log Periodic unit.

INSTALLATION NOTES

Height of Antenna. TV signals have a short wavelength and thus travel in a straight line. See Figure 3-1. Place your antenna as high as you can to make sure it captures a distant signal.

If you are brave enough to fight with an outdoor antenna and mount it onto your roof, see the book *The Right Antenna* by Alvis J. Evans for reference. In the meantime, flip to Chapter 10 in this book for more information on hardware, accessories and hookup procedures for antennas.

CABLE TV

There are hundreds of channels of infomercials, reruns, politics, etc. We love having all of these channels and can't live without them. In short, this is cable TV. Sixty-five million homes in America are wired directly to this revolution in television broadcasting. How does it all work, and what do you need to know to improve your television viewing via cable?

Cable Kicks Off. Cable had its origin in a scheme known as community antenna TV (CATV). During the time of CATV, if you lived in an area where TV signals were weak, such as in a valley or far from a broadcasting antenna, your TV's antenna would not pick up much of a signal. This created a bad picture, or snow. The solution was to have one huge antenna hooked to multiple signal amplifiers. Then the strengthened signal was split and sent to several homes via a cable. This collective system allowed users to share the cost of the equipment instead of having each person using a huge antenna.

In 1975, RCA took the growing CATV market and went one step further. Cable TV was born with the launch of RCA's first satellite. Ground-based cable companies would now purchase satellite equipment, pick up RCA's signal and resell the programming to local residents via the CATV network. Cable companies now obtain signals from several sources, such as powerful antennas and numerous satellites. They basically became rebroadcasters or middlemen: information peddlers.

Signal. Most major cities and surrounding areas in the U.S. and Canada have cable access: miles of coaxial lines, helped by amplifiers and cable support personnel.

Receiving cable TV is elementary. You call the local cable company (there is usually only one company servicing your area) and have them install the lines from the street to your TV or TV-top cable boxes. In some cases, renting a cable box is part of the unavoidable process. The hardest decision for many people is still whether to order HBO or The Movie Channel.

Signal Quality. Cable's signal quality is susceptible to common analog problems. Amplifiers in the cable loop go down, cables can be accidentally cut or frayed by workers, etc. As a result, cable is not known for its reliable signal or quality. A cable signal is typically of lower quality than even an antenna signal, let alone DBS. However, some cable companies do offer stereo and surround sound.

Signal Choices. Each cable company provides certain channels. They basically sell you *their* signal picks from around the world. Many factors affect what channels are carried. If your cable company is missing a station, call and ask them to carry it. If they refuse (or laugh!), maybe it is time to look at a DBS system. Cable no longer has a monopoly on quality or quantity broadcasts.

Signal Scrambling. Due to *pirates*, cable companies are forced to use different kinds of scrambling or signal trapping techniques. The most popular method is to scramble the signal from its source, then require everyone in the cable company's network to purchase or lease cable descrambler (decoder) boxes. This creates some serious remote control and VCR paradoxes. See Chapter 10 for different hookup schemes you need to use for these descramblers.

Another method of scrambling is for the cable company to install special electronic signal traps just outside your home. When you order a certain channel, they simply come and remove the trap for it. Another scheme companies use to disable pirate boxes is called a *cable bullet*. They send a pulse down the line which disables your decoder box and essentially turns it into a paperweight.

Cable Boxes (Receivers), Cable-Ready TV and VCR Tuners. Cable boxes are needed to descramble or simply receive a cable signal then convert it. Most TVs and VCRs sold today have one of these units built right in. Keep this in mind when shopping for a new TV or VCR. Purchasing this built-in cable ready feature can save you from equipment duplication.

Beware! You may be able to receive basic cable channels, but will have to use a cable company's decoder to receive premium channels. This depends on the scrambling method used. Contact your local cable company to find out which channels can be received with a cable-ready TV or VCR.

Also beware of the fact that some manufacturers advertise TVs or VCRs that receive 181 channels. This doesn't necessarily mean 181 cable channels. In other words, it may not be cable-ready just because it says 181 channels.

Pay-Per-View and Video On Demand. For those of you who have a cable box, you can "buy" movies and events broadcast on pay-per-view channels. When you pay to view a movie, the channel's signal is descrambled when the movie starts and rescrambled when finished. Costs of movies are typically $3 to $4 each, with adult movies just slightly higher. Sporting events range from $5 all the way to $75+.

Picking a Cable Company and Dealing With Them. Unfortunately, most cities only have one cable company that services your specific area. This creates a very bad monopolization situation. Cable is infamous for its unreliability, high costs, outages, signal noise and multiple price hikes. When you run into problems with a cable company, all I can recommend is to complain to every consumer group you can and make sure you tell everyone to stay away from the company. However, if you find a very service-oriented company, boast of them to your friends. The last resort of dealing with cable, and one people are taking more often, is to buy a small satellite system and service. The quality and reliability of DBS's digital signal is heaven-sent (literally).

Content. In addition to local channels (which are the most-watched channels in most households), cable offers such quality programming as sports, news, general audience, government access, children's shows, music, super stations, home shopping networks, rerun networks, science, and movie channels. Pick up a local *TV Guide* or newspaper for a complete list of channels available in your area.

Cost. Cable is usually sold in service packages:

Basic Service: Usually twenty or so channels, most of which are local broadcasts. Cost: $15 to $20 per month.

Expanded or Plus Service: Includes basic service plus a few channels that few people can't do without. Most people choose this option as it is only a few dollars more than the basic service. Cost: $25 to $30 per month.

Movie Channels or Pay TV: These are usually packaged, and range from $12 for one channel to $20 for three.

Pay-Per-View: Typically, movies are $3 to $4 each, with adult films $6 to $8 each. Sports shows are widely varied.

Equipment Rentals: Be prepared to pay a hefty deposit. Converter/descramblers with remotes are about $6 to $10 per month.

Installation Fees: If you're lucky, the company is running a free installation special. If not, expect to pay $20 for a house that already has the cables, and up to $100 for one that requires "the works." Additional outlets and converters will cost more.

Converter Features. Some of the features include volume control, VCR filter, VCR timer, full TV-top keypad display, IR wireless remote, parental control, channel recall, stereo TV compatible, and pay-per-view.

Future. Cable's future is "up in the air" at the moment of this writing. Should cable go digital, with 500 channels or HDTV? It would cost each customer $500 in upgrade costs. Who would pay that? Should cable companies raise the number of channels and leave the quality low? This is a more likely choice. Services such as video-on-demand and high-speed Internet access would be possible.

Wireless cable? An oxymoron if I ever heard one. The recent Telecommunications Act of 1996, passed by the FCC, states, "The goal of this new law is to let anyone enter any communications business; to let any communications business compete in any market against any other." This means phone companies can provide cable services and cable companies can provide phone services. Enter wireless cable. It is sent over high-frequency telephone network towers. Almost a million people currently subscribe to wireless cable services. The phone companies are putting up networks as I speak. The Telecommunications Act now allows cable companies to deliver high-speed Internet access and other telephone options to customers.

In the future, there will be a fiber optic digital network that will send and receive every bit and byte of information into our homes through one hair-sized glass fiber. 'Til then, we'll keep paying a fortune for low-quality, high-quantity programming.

SATELLITE DISHES

A satellite (sometimes affectionately referred to as a "bird") is a communications device placed in Earth's orbit. It receives a signal from Earth then sends it back to a vast surface area. It basically *relays* information. A satellite receives powerful high-frequency FM (frequency modulated) broadcast signals from dish-like transmitting antennas (uplink transmitters), converts them to a different frequency, and sends them back to Earth as low-power signals. People can pick up the signals with a similar dish-like antenna and signal amplifier. From there, a satellite

NOTE: The signals are converted at the satellite to prevent incoming and outgoing signals from interfering with each other.

receiver is used to translate the signals into visible pictures on your TV, and sound through your speakers.

Geosynchronous Orbit. Also know as *geostationary* or *fixed orbit*. Satellites are place in a geosynchronous orbit 22,279 miles above the equator. See Figure 3-2. To give you an idea how far away this is, the diameter of the Earth is only about 8,000 miles, and the circumference is 24,902 miles at the equator.

This means that if you point to the satellite from Earth, it will not move from that position over the equator. Ever! The satellite spins along with the Earth, powered by its own inertial forces. This is so we can aim our satellite dishes at a given spot in the sky and pick up a signal from a specific satellite.

The Signal and Its Path. See Figure 3-3. A signal originates from a satellite uplink (ground-based transmitter). This can be a TV station's satellite dish or even a portable device placed atop a van. It reaches one

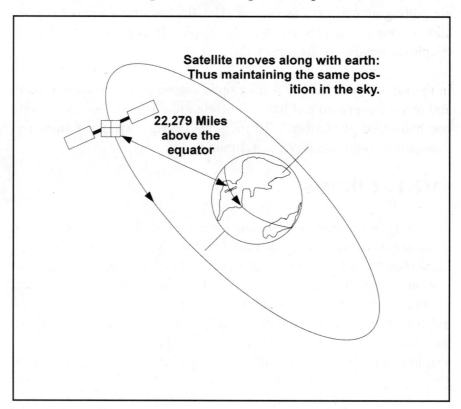

Figure 3-2. Geosynchronous satellite orbit.

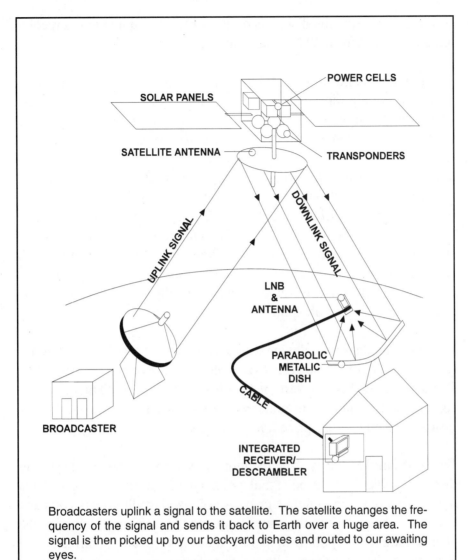

Figure 3-3. The workings of a broadcast satellite.

of the current satellites placed in geosynchronous orbit. Once received, the TV signal is then amplified and converted into different frequencies. The satellite itself carries a very small power supply to send a fairly weak signal back to Earth. A standard satellite typically uses 8 to 16 watts. DBS uses 120 watts.

The satellite's signal is then beamed back to Earth to be picked up by the satellite antenna in your backyard. When the signal is finally received by your satellite dish, it is quite feeble: the equivalent of receiving a CB signal from 22,000 miles away.

Footprint. The Earth-area of the satellite's coverage is called its *footprint*. If you are more centrally located in a satellite's footprint, you will receive a stronger signal. The farther out you are, the more amplification is needed and the bigger the size of dish required.

The Dish. Because the satellite's signal is weak, a large antenna is needed to receive it on the ground. The antenna is actually a large parabolic (cross-section) circular dish. The dish is pointed at a satellite 22,279 miles above the Earth to receive a microwave signal. The weak signal transmitted from the satellite strikes the entire surface of the dish and is focused onto one point called the *feedhorn*. In other words, the dish takes a large surface area of signal (typically 6' to 12' in diameter) and packs it into a small spot.

NOTE: See the DBS section for information on DSS-type 18" dishes.

Inside this feedhorn is a device called a *low-noise block* downconverter (LNB). It amplifies the signal received by the satellite by more than 100,000 times, then converts it to a lower frequency. This is so the signals can be sent to your home with a standard RU-6U coaxial cable instead of a huge, high-priced, high-frequency cable.

Feedhorn and Polarity. The feedhorn is positioned above the satellite dish to receive the focused signal. A group of concentric rings selects the *polarity* of the incoming signal. A polarized feedhorn lets you pack twice the channels into the same frequency by receiving two signals. One is horizontally polarized and the other is vertically polarized. See Figure 3-4. The feedhorn tells which polarization to accept for which channels. Odd channels have one polarity, and even channels have the other. Ingenious!

Frequencies and Transponders. Satellites operate in the microwave frequency range (1 GHz to 300 GHz). This is much higher than the radio waves used back on Earth for television broadcasts. The reason such high frequency microwaves are used is because they can be focused very well over long distances, and are more resistant to noise.

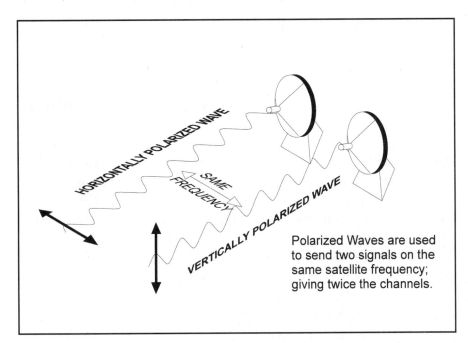

Figure 3-4. Polarization in satellite dishes.

A satellite typically carries several transponders. The transponders take the high-frequency microwaves received from Earth, amplify and convert them to a slightly lower frequency, and send them back to Earth. Satellites are classified by the frequency band (of microwaves) that the onboard transponders use:

C-Band: The uplink uses frequencies in the 6 GHz (6,000,000,000 Hz). The satellite converts this to a downlink signal of around 4 GHz. There are typically 24 to 54 transponders on each satellite, each using 5 to 10 watts. The low power makes it necessary to use a large dish to receive the transponder's signals on Earth. Just a note: both C and K-band units are analog, compared to the new digital Ku-band birds.

K-Band: The uplink is in the 14 GHz range, and the downlink is around 12 GHz. The power used for each transponder is much higher, around 40-50 watts per transponder; therefore allowing the satellite to only carry 10 or so transponders. The advantage is that the ground-based dishes need to be only 2' to 3' in diameter. Because the frequencies used are much higher on the microwave band, they are less susceptible to noise.

NOTE: There are hybrid satellites that mix C-band transponders and K-band units onto one satellite.

Chapter 3: Receiving With Antennas, Satellite Dishes & Cable

DBS satellites will be described later.

C-BAND SATELLITE EQUIPMENT

Here is a list of what you need to receive signals from a C-band satellite:

Dish. C-band satellites require a large dish for receiving a signal. The dish is made of metal or metal mesh. Dishes come in various configurations and sizes. Size is more a matter of where you live in the satellite's footprint, and configuration is a matter of preference. If you live in a high-wind or high-snow area, use a metal mesh dish to prevent the poor thing from being blown over or filled with snow. They also tend to blend into the landscape better, as well as any 12' disk on a pole could. A black anodized aluminum mesh unit is a popular choice. You may choose a steel, aluminum or fiberglass coated solid dish. The fiberglass units are actually fiberglass coated onto a metal surface. All of these dish types are durable and sturdy.

LNB. These are rated in *gain* (amplification) and *noise temperature*. The higher the gain, the better: 50 to 60 dB is quite good. The noise temperature rating is a measure of the "cleanliness" of the output of the LNB. This is rated in Kelvin. Look for a 25 degree or less unit: the lower, the better.

Motorized Mountings. Because there are multiple satellites, you will need a mounting that lets the dish *pan* the sky. On a motorized mounting, a motor is used to turn a screw rod or rotor system which rotates the dish. You simply install it and aim the dish right from your easy chair via remote control. Try to get a unit that goes from horizon to horizon so you can pick up every available satellite signal.

Receiver. This is the TV-top box which converts the signals from the LNB into TV signals. It also controls the polarity in the feedhorn, controls the motorized mounting to aim it, and descrambles channels. Look for an IRD (integrated receiver/descrambler) and not just a receiver.

Descramblers. Some satellite channels are sent as scrambled signals. You must have a receiver with a built-in descrambler to receive them. Also, you have to subscribe to a monthly service such as Video-Cipher II and pay the fee to receive scrambled channels.

Advantages/Disadvantages of C-Band Satellites. Costs of satellite systems are another downfall. You can easily spend thousands of dollars. Are a few free channels worth it? Small, bird feeder-sized DBS satellite dishes have all but taken over the market. However, don't hold the funeral for C-band birds yet. You can still pick up quite a few channels, audio programs and live feeds with a full-size dish. You can also purchase a cheap, used system from someone upgrading to DBS.

DBS

Direct Broadcast Satellite and Other Small Ku-Band Satellite Systems. In 1994, people revolted against the cable companies, and the choice of a high-quality digital product over the poor and problematic analog cable services was born. Cable's media monopolization ended. The most successful electronics product to ever hit the market was launched that year: small satellite dishes (a.k.a. direct broadcast satellite (DBS) or Digital Satellite System (DSS). DBS is the blanket term we use to classify all small satellite antenna systems and services.

DSS is a trademark owned by Hughes Electronics Corp. It has come to be somewhat of a generic term for direct broadcast satellite technology. Other companies, such as PrimeStar, offer similar DBS technology and use medium-power Ku-band satellite technology. A few other companies have services similar to DSS.

Overview. DBS is basically a cable replacement. It offers 175+ channels of programming that you receive with a new generation of micro-satellite dish antennas: a cheaper and more compact version of the current big boys found in the back yards of houses and the Motel 6 down the street. Currently, the diameter of these dishes is between 18" and 40"; a mere Frisbee compared to the 12' flying saucer-sized C-band dishes.

CABLE vs. DBS

Which is better, cable or direct broadcast satellites? Let's take a look at the stats and pros/cons of each:

STATISTICS	CABLE	DBS
Horizontal Resolution	-300	400+
Sound	Mono or Stereo	Surround Sound (both)
Customers	65 Million	5 Million
Hardware Costs	Lease box for $6-10/month	$100-500 for equipment or you can lease.
Installation Charges	$20-100	$100-200 or free with lease.
Programming Costs	$15-20 Basic $25-30 Expanded	$15 Basic $25-50 Expanded
Antenna Needed for Local Channels?	No	Yes
Hardware and Service Reliability	Depends on the area	Sometimes mid-high but is generally low
Signal	Analog	Digital/Compressed

ADVANTAGES/DISADVANTAGES

Cable. Cable generally has high-quality programming, and in some cases is cheaper than DBS, but is severely prone to outages and weak signals. The programming is overpriced for the low-quality signal and service you receive.

DBS. DBS has 30% better picture resolution than cable and is 65% sharper than a VHS movie, but is occasionally prone to interference and has a high start-up cost for equipment and installation. Programming costs are nearly equal to cable, but you get a better picture for your money. The other disadvantage is that you need an antenna to watch local channels. Also, DBS's digital nature makes for a stronger noise-to-signal ratio: you either receive a pristine picture or no picture at all.

You need a DSS satellite dish (or equivalent) and a TV-top integrated receiver/decoder (IRD) to receive programming. The IRD is the box that you connect to your TV. Add to this a remote control and a programming access card. The cost of one of these units is currently hovering around $399 to $850 without installation. Some DBS services are offering equipment rebates with the purchase of one year's worth of programming subscriptions. Yes, the down side to this mass media machine is that *you have to pay a monthly fee for programming*. Expect to pay about $15 a month for basic channels: maybe $60 a month by the time the video reality dust settles. Contact each company for current pricing, packaging and equipment rebate information.

DBS services such as DirecTV, USSB and PrimeStar are basic equivalents to cable companies. They uplink the broadcasting channels to the satellites for distribution back to your home. This is called direct-to-home service (DTH), or DBS service.

Equipment and Costs. Several companies are hatching this new breed of bird (satellite dish). Most offer the dish, receiver and service as a package deal. Here is a quick rundown of what you will need and what is currently available. Contact your local services to get current package deals:

See Figure 3-5. A typical DSS system consists of an 18" dish, digital TV-top decoder box, remote control and access card (the access card is used to receive pay-per-view movies and events). PrimeStar, DISH Network and other DBS companies have similar packages but use varying-sized dishes.

Hardware Manufacturers: For the DSS system, manufacturers include Daewoo (Eurosat), Fisher, GE, Hitachi and Memcorp (Memorex),

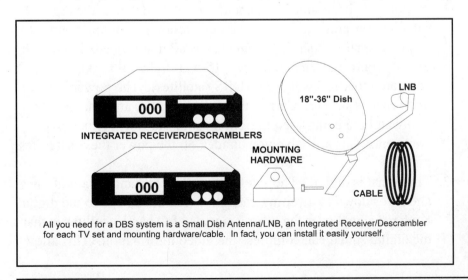

Figure 3-5. Direct broadcast satellite components.

Chapter 3: Receiving With Antennas, Satellite Dishes & Cable

Hughes Network Systems, Magnavox, Matsushita (Panasonic), Philips, ProScan, RCA, Samsung, Sanyo, Sony, Toshiba and Uniden. All copyrights and registered trademarks are owned by their respective companies.

For other DBS satellite equipment and services, manufacturers include AlphaStar, EchoStar (DISH Network) and PrimeStar. With some of these services you have to lease *their* DBS equipment.

Each company has several models and feature lists: one unit for receiving the satellite info, and additional IRDs for other TVs in the house. Check out their Websites or contact each company for further information. See Appendix A for a list of Website addresses for these companies.

Technology. There are *three main differences* between DBS and full-sized satellites:
1. For DBS, high-powered transponders aboard satellites run the show from space. They use a much higher wattage and higher frequency to send out their signal. This means less interference and signal degradation. Just a note: Pictures are not weakened with a DBS system. You either receive the digital signal and a pristine picture or no picture at all. Hughes Electronics has launched three of these DBS satellites, which feature sixteen 120-watt transponders (receiver-transmitter). They make use of the Ku-band. Older satellites (C-band units) use about 16 watts. Because the digital satellite transponders are putting out a relatively large amount of power, a smaller dish (18" diameter) is all that is needed to pick it up on terra firma. Just imagine a 15-watt light bulb (old satellites) compared to a 120 watt bulb (DBS satellites). The power difference is similar. Carry this analogy further and imagine this 120-watt bulb being used to light most of the North American continent. This will give you an idea of the power these satellites put out.
2. The signal is digital. The satellite sends a constant stream of ones and zeros down to awaiting gadgetry. The advantage to the digital Ku-band signal over the analog signals of C-band satellites is that the digital signal can compress the video information. This allows

for many more channels and services to be packed into each of the onboard transponders.
3. The overall audio and video quality from the digital signal is currently unmatched and astounding. The unit uses modern data-compression technology to deliver CD-quality sound and 400+ lines of video resolution. Cable lacks this quality and DTV is still on the drawing boards.

Installation. The small DBS dishes are a dream to install compared to the incubus task of planting a full-size satellite dish in your yard. You can either hire a service person (for about $150 to $200) to handle the installation fiddling and fumbling, or mount the small dish yourself (about $70 for a kit). Installation is performed in this order. Make sure to read and follow the manufacturer's instructions:
1. Assemble the dish antenna and LNB.
2. Mount the dish on the side or roof of your home or somewhere in your backyard. A mounting kit may be needed at an additional cost. Looking for a clear line of sight to the satellite in the southern sky (101 degrees west longitude position.)
3. Ground the dish.
4. Install and route the cables.
5. Hook up the IRD (black box).
6. Point the dish with the signal strength feature. This is done by inputting your zip code then adjusting the dish's *azimuth* and *elevation* until the on-screen signal strength meter says all is fine.
7. Hook up the phone lines if applicable.
8. Order your service and card.
9. Enjoy!

DBS Services. As mentioned before, you will need a DBS service to receive programming. Here are your options:

DirecTV: This is a subsidiary of Hughes Electronics Corp. It will supply your new DSS system with 175+ channels, cheap pay-per-view channels, and 31+ music channels. Access cards and connections to phone lines for billing and pay-per-view may be necessary. Website: http://www.directv.com. Phone: 1-800-DIRECTV. *Costs per month*: $15 for about 20 video channels and 35 music channels all the way to $48 for

NOTE: Some DBS units, such as Sony's, have a built-in LED signal strength meter right on the dish itself. This saves you from having to run up and down or lose your voice yelling to a helper.

NOTE: Some areas of the country have ordinances regarding the installation of dishes under one meter in diameter on your property. Ask the DSS dealer, DBS service or the FCC for more information. Also note that some DBS services will install the dish for you as part of the package.

Chapter 3: Receiving With Antennas, Satellite Dishes & Cable

about 90+ video channels and 31 music channels. Pay-per-view events are extra.

USSB: United States Satellite Broadcasting. A DirecTV licensee. This adds another 22 channels to your DirecTV service, for $7.95 to $34.95 a month.

EchoStar (DISH Network): This service uses a high-power satellite network (like DSS's) and is very similar to DirecTV except you need a dish network or HTS-produced 18" dish. Both EchoStar and DirecTV offer an interactive programming guide to manage the mass amounts of channels. Website: http://www.echostar.com. Phone: 1-800-333-DISH. *Costs per month*: $10 to $40

AlphaStar: This system uses a 30" dish. They offer cheap equipment and service, but the equipment is somewhat scarce at this time. *Costs per month*: $25 to $50

PrimeStar: PrimeStar is promoted as being a turnkey package. They provide the leased equipment, install it, and feed your new hungry dish its digital meal. Currently there are 160+ channels. PrimeStar uses a large dish (39" diameter), as its satellite transponders are not as powerful as Hughes'. Contact PrimeStar or Radio Shack for further information. Website: http://www.primestar.com. Phone: 1-800-PRIMESTAR. *(The weird phone number is not a misprint.) Costs per month*: An installation fee may apply to this service. Programming costs are $33 to $55, and $10 for the equipment rental.

Systems are not compatible with each other. So if you decide to switch services, be prepared to start over with new equipment costs as well. If you want to know which system is popular in your area, pick up a newspaper and look at the classifieds. The unit that is least resold is the favored one.

Sorry, northern neighbors; Canada is DSS and DBS no-man's land. They are not currently offered in your provinces due to broadcast legalities. However, networks and services are working together to provide Canada with service soon.

On-Screen Menu Systems. Each IRD has a built-in on-screen menu program. Programming information is fed to it by your DBS service. DSS systems allow you to browse program listings up to three days in advance and call up a description of each listing with one click. EchoStar and AlphaStar require you to step through several menus to get the info. PrimeStar has a simple, constantly scrolling grid offering no ability to interact with it.

FUTURE OF SATELLITES

The battle between cable and DBS will determine the future, but here are some facts to help you with your buying decisions:

DTV and Wide-Band Data Port. Some DSS systems have a built-in port that you will be able to use when the new DTV standards are reality. Check with manufacturers to see who offers this feature. If purchasing a new TV-top box, consider buying one with this feature so you will then be able to make use of DTV's proposed high-quality sound and resolution when it arrives.

Narrow-Beam Transponder Technology. In the future, satellites may be equipped with a *spot beam*. This would allow a more localized footprint so services can offer programming that is more specific to your area, including high-speed Internet access and local channels.

Built-In Decoders/Receivers. Once the battle settles a bit, TV and VCR manufacturers may begin to offer built-in IRDs for DBS systems. Keep this in mind and save some money when these units are released.

Internet Downlinks. Several companies are working on using satellite networks to provide data downlinks to your computer. This would work in conjunction with a phone modem in order to bring you stomach-churning, high-speed web surfing and Internet services. Also in the planning stages are pay-per-computer programs or pay-per-program-use schemes. A satellite would simply download a program to your computer via your dish, and you would pay for each use of the whole program.

SUMMARY

A well-thought-out plan to receive broadcast signals is essential to your home entertainment package. My own experiences, and the stories people have told me, compel me to air my opinion here (for whatever it's worth). I have been swayed by DBS systems simply because of the quality and past nightmares of dealing with cable companies (who may improve as the technology changes). The final decision is yours, though. As far as antennas go, I would definitely purchase a good Log Periodic model even if you have cable or DBS. This is the best simple upgrade you can make to your signal receiving system. It's strictly your decision on the other dilemma: cable or DBS?

CHAPTER 4

VCRs, LASER DISK PLAYERS & DIGITAL VIDEO DISK PLAYERS

"My VCR flashes 01:35, 01:35, 01:35, ..." Steven Wright

Cable, satellite systems and antennas let us receive the broadcast signals that entertain us. The information that comes through these signals operates on a given schedule, meaning you are forced to watch what the local broadcast, cable and satellite companies want you to watch. However, what if you want to see a new blockbuster movie that was just released? What if you want just to sit with your mate and enjoy a romantic classic flick? People want to watch movies and programs *THEY* pick, when *THEY* want. The solution is to play a prerecorded feature on a VCR or video disk player.

Also, what if you want to record a program that is on during the day so you can watch it after work? VCRs allow you to indulge in this "delayed broadcast." See Photo 4-1.

In this chapter, we will look at how VCRs operate, what advances are being made, and the new video disk technology which may take over the VCR's reign over the video realm. Let's glimpse the future of this video media.

Photo 4-1. A four-head hi-fi stereo VCR. Reproduced with the permission of Sharp Electronics Corporation.

Chapter 4: VCRs, Laser Disk Players & Digital Video Disk Players

Types of Recording and Playback Media. There are two basic forms of media for watching movies, and one for recording programming. They will be described later. See Figure 4-1.

VIDEO CASSETTE RECORDERS (VCRs)

Basics. As a simple explanation, a video cassette recorder is an electro-mechanical device that records and plays back video and audio signals. How about a jazzier definition? A VCR is a portable magnetic media record/playback device that feeds our eyes and ears spectacular scenes and songs. VCRs utilize spooled cassette tapes of various sizes depending on the format. This is the *magnetic media* referred to previously. The videotape is actually a metallic material coated onto a plastic tape upon which the VCR magnetically deposits an analog TV signal, or in some cases a digital one. The VCR *records* using a pint-size electromagnet embedded in the VCR *heads*. The signal is then retrieved for playback using the same heads. Once the signal has been read off the tape, it is amplified, cleaned and delivered to your TV via a cable or set of A/V lines. Voila! A prerecorded movie appears.

History. For years, broadcast stations used sophisticated 3/4" magnetic record/playback devices. However, an average person couldn't afford one. Screams of technological emotion where heard in 1975

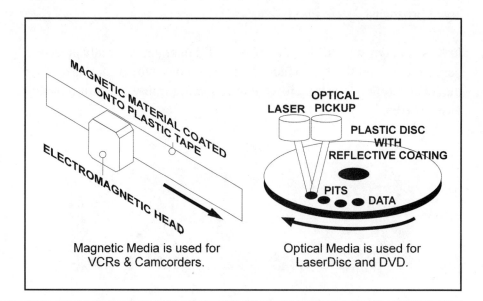

Figure 4-1. Types of recording and playback media.

when the first VCRs were commercially accepted by the public: Sony's infamous 1/2" *BetaMax* (3/4" and 1/2" indicate the width of the videotape). Clunky, overpriced, unreliable behemoths, but we adored them!

JVC decided to fight the growing Beta craze with its own format, VHS. Their engineers were actually ordered to design a device that would meet the absolute "minimum" video quality the public would accept. Sort of like designing a crayon instead of a precision pen. However, the crayon was cheap, worked well, and was therefore accepted as the de facto standard for VCRs.

VCR Uses. The original whoopdedoo over Beta was because consumers could finally record their favorite shows for later viewing, without paying $10,000 for a recorder. This was the new VCR's major selling point, but other purposes soon evolved.

What are VCRs used for these days? Most people simply hook them to a TV, rent an overpriced "new video release," then give the well-worn *play* button another workout. Fearless souls may even attempt to assimilate a VCR manual and brave the dust-coated *record* button, even without a child's help. This is no longer a timorous task as new features like VCRPlus+ and StarSight make recording your TV shows child's play (no offense to adults).

What else can be done with VCRs? Video dubbing: recording one tape onto another. Most mid- to high-range VCRs have built-in stereo decoders as well as cable-TV receivers. This is great when you want to save money by preventing equipment duplication. An enterprising individual even figured out a scheme to use a VCR as a digital data depot for computers (tape backup unit). Whatever you do with your VCR, it pays to spend a few extra bucks on features such as hi-fi and four heads now, and avoid rebuying later.

BASIC COMPONENTS UNDER THE HOOD

Have you ever seen the inside of a modern VCR? You may be surprised by its mild-mannered, simplistic innards. Don't be fooled. It is still a superhuman electromechanical video marvel. Here is the basic breakdown of a VCR:

Mechanical Components. These include the motors, belts and support linkages which crank the tape around its path. Most of the gears, arms, doohickeys and thingamajigs are plastic. With age and use, they wear out and become repair shop bait.

Heads. A *head* is a miniature electromagnet encased in a metal cylinder. When it is energized, and the metallic-coated VCR tape passes over it, that spot on the tape becomes magnetized (recorded onto). Later, the magnetic data is used to play back the originally-recorded video signal, over and over if necessary. The typical VCR contains two, four or more magnetic read/record video heads, an audio head, an erase head, and a synchronization head (timing). Each VCR format and model has different head configurations. Refer to Figure 4-2.

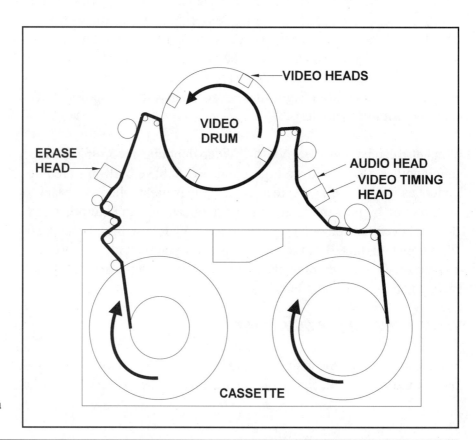

Figure 4-2. The heads in a VCR.

Video Heads: These pinhead-sized electromagnets (usually two to four) surround a rotating drum. They are used to magnetize the metallic material contained on the VCR tape. See *How the Drum Works* for further information.

Audio Head: The audio head is separate from the video heads. It lays an analog sound track along the top of the videotape. Audio takes up much less space than video, the tape real-estate hog. The width of the soundtrack is .7 millimeters, which is much smaller than a comparable audio tape. For a normal mono recording, this space is sufficient.

Hi-Fi Stereo Audio Heads: Hi-fi needs a much larger space than the pitifully small .7 mm mono audio track. Therefore, manufacturers use two audio heads embedded into the video head drum. See Figure 4-3. These special heads record the audio signal first, on a *deeper layer* of the tape (because there is no room between video tracks). This allows the video signal to be recorded right over the top of the hi-fi signal. The mechanics of this are complicated and beyond this book, but the results are stunning. Nothing like sweet-sounding stereo and a magical hiding place to pack it!

Erase Heads: These two heads *erase* unwanted audio, video, and timing information from the tape. One of them is the first head the video-

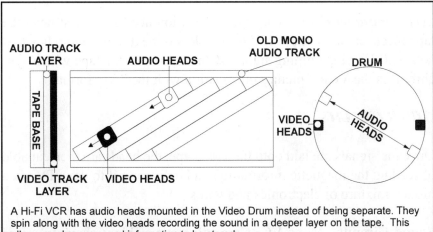

Figure 4-3. How a hi-fi VCR records sound.

Chapter 4: VCRs, Laser Disk Players & Digital Video Disk Players

> ## HOW THE VIDEO DRUM WORKS IN A VCR
>
> Video uses scads more magnetic tape real-estate than audio. Therefore, a VCR needs a way to read more data in a given time span. It does this by moving heads over the tape much faster. The actual tape speed does not increase; the VCR instead uses a gyrational trick.
>
> The actual tape is only moving forward at a few inches a second. However, by mounting two, four or more video heads into a tilting, rotating drum, the VCR simulates a 250"-per-second motion rate. This is fast enough to pack the needed video track onto a small 1/2" tape.
>
> The angled drum rotates 30 times a second, creating diagonal strips that contain the magnetic information. This is called *helical scan*. The drum contains a minimum of two heads. One head reads the odd field 30 times a second, and the other reads the even field 30 times a second. In this way, the video heads match the TV's NTSC standard of 30 frames a second. Very slick trick!

tape meets on its journey. Like an electromagnetic rubber eraser, it will erase all of the audio, video and sync information at once so you can record onto a squeaky-clean tape. The second head is a separate unit positioned in proximity to the audio head. It allows you to record a whole new soundtrack but keep the same video tracks. For information on *flying erase heads*, see Chapter 9.

Synchronization Head: The synchronization head is the last head the tape meets on its journey. It is right below or next to the audio head. It lays a 30-Hz video timing signal on the bottom of the tape which synchronizes the VCR's picture information with the TV's format.

OTHER CIRCUITS

Once the signals are laid onto the videotape, the heads are then capable of reading the magnetic information and delivering it to your TV via a motley mixture of electronics and wires.

Sound Circuits. Mono VCRs are still made. This is shameful. It is well worth the extra $50 for quality built-in hi-fi stereo sound circuits.

If you are going to expand your home theater, it's necessary to own a hi-fi VCR. Also, if your TV doesn't contain an MTS decoder, you may want to add this feature to the VCR list of goodies.

Tuner Circuits. VCRs don't need to be hooked to a TV to record your favorite show. They have built-in television tuners. Most have a built-in cable TV tuner as well. This allows you to surf or videotape your cable company's video pickings as well as hot local broadcasts. Look for a unit that has a cable TV tuner and an automatic channel scanner. Also, find a VCR that automatically locks onto and programs cable stations (plug-and-play). It may save you the leasing price of a black box from your cable company. Beware, however: some cable services won't allow the use of a built-in cable TV tuner. You might have to record and view the signal from the cable company's TV-top cable box.

Remote Circuits. All VCRs come with either an elaborate zillion-button remote control or a super-simple one-button toy. Some even feature on-screen programming and VCR Plus+. Look for one that comes with a universal remote or a code-entry remote. These control your TV, cable box, VCR and whatever home theater equipment you have. It also saves you from spending $15 later on a universal remote. See chapters 2 and 9 for further remote control features and suggestions.

Programming Circuits. Older VCRs have complex external button sequences that even a navy radio technician would have trouble deciphering. On-screen menu-assisted programming is taking away the ingrained fear of recording your favorite TV shows. Simply use your remote to fill in the blanks on the screen, such as the time the show starts and how long the VCR will record. Some cable boxes even interface with the VCR to make recording programs easier.

VCRPlus+: This is one of the great inventions of the video age. Better even than the wonder of the VCR itself. VCRPlus+ is basically an automated programming feature. Next to each show listing in the local *TV Guide*, there is a stream of numbers. By simply tapping these numbers into the VCR, the machine knows when a show is on and how long to record it. See Photo 4-2.

Photo 4-2. A VHS hi-fi stereo tri-logic VCR with VCRPlus+, digital auto tracking, auto head cleaner and flying erase head. Reproduced with the permission of Sony of Canada Ltd.

Input/Outputs. All VCRs have an input/output line for a modulated signal (the coaxial hookups are called RF connectors), and a set of A/V jacks (cinch) for composite video and audio. If you are going to transfer video from your camcorder to a VCR, you may want a VCR with front A/V jacks for convenient hookup.

New VCRs come with an S-connection (also called S-VHS or Y/C) which splits the luminance and chrominance. SCART (A/V Euroconnector) and other similar units are special multi-wired connectors that control many devices at once. This leaves room to expand your system for future digital standards. For an expanded list of features, see Chapter 10.

TYPES OF VCRs

VCRs are basically categorized by the tape format they wield. BetaMax was the beginning, but soon died a painful death. VHS and VHS hi-fi have remained the strongest contenders. Most movie stores only carry the VHS format of prerecorded feature. Now others have joined the battle: Super VHS, S-VHS-C, VHS-C and 8-mm.

Video Quality. Standard VHS is inherently low quality. In fact, it is about 25% worse than a broadcast or cable video signal. We're talking about horizontal resolution.

Super VHS is of superior resolution, about 66% better than standard VHS. Eight millimeter (8-mm) is similar, but not many movies are in S-VHS or 8-mm. So, should you buy one of these VCRs? Only if you will be recording DBS signals or using it to edit from an S-VHS or 8-mm camcorder. Otherwise, save your money until more movies are released in these formats.

On-Screen Programming. The new VCRs are able to send a signal to your television so you can use its on-screen programming features. This is really handy if you are recording while parked in the easy chair.

EDITING WITH A VCR

If you will be editing your home movies with a VCR, try to find one with options geared toward the job. Most manufacturers have models that are designated as editing VCRs. Otherwise, look for a jog/shuttle feature and, if possible, a flying erase head.

WHAT TO LOOK FOR IN A VCR: RECOMMENDATIONS

Four Heads. You will see a VCR advertised as a two-head or four-head model. This refers to the number of video heads embedded into the video drum. The reason why there are four heads instead of two on newer models is because four heads allow for better special effects, such as freeze frame and slow motion or frame advance. The cost difference is fast becoming negligible, so always opt for four heads.

Hi-Fi Stereo. Quality sound circuits have made their way into VCRs as well as televisions. A high-fidelity VCR allows you to record and play back near-perfect sound. As with four-head VCRs, the cost between mono and hi-fi stereo is next to nothing these days, so go for the hi-fi and enjoy the superior sound quality.

Plug-and-Play. Almost every new VCR, even a low-cost model, has this feature. It lets the VCR program the time and set the cable channels on its own.

Simple On-Screen Programming and VCRPlus+. A mid- to high-quality VCR will include these features. If you will be doing a lot of recording, make sure you have on-screen programming and VCRPlus+ on the VCR.

Front Jacks. These are A/V jacks placed on the front panel of the VCR. They let you easily plug your camcorder into your VCR for video editing. See Chapter 9 for more VCR features.

Video Cassette Recorders vs. Video Cassette Players. The price difference is generally negligible between these machines. Besides, what happens the first time your TV viewing plans coincide with the dinner reservations that you made with your significant other? The fact your VCR records may save your relationship; worthy of an extra fifty bucks.

VCR/TV Combo. I hate to be a devil's advocate for this awful machine, though there are one or two uses for this invention. The fact is that the two components (TV and VCR) are just as cheap separately, and just as light. Another disadvantage is, if one of the elements (the TV or the VCR) breaks, then you are stuck with having to purchase a whole new unit at twice the price of only having to replace one function.

On the other hand, a VCR/TV combo may be the only means to civilized sanity out in the middle of nowhere. In other words, you may want to use it in a camper. The other use is in a kitchen or in a company meeting room, to help save on clutter. Prices are around $350 to $700.

VCR REPAIRS

VCRs have fallen victim to the disposable device mentality. Integrated electronics and robotic mass-manufacturing processes have brought VCRs down to the sub-$200 level. This is a far cry from the $1,200 "investment in the future" of past models. It is to the point where it is cheaper to replace a VCR than to send it to a shop for even the simplest mending. However, you can save yourself a little cash and heartache by learning how to clean and perform minor repairs on your own VCR. See Chapter 10 for more information, or read *The In-Home VCR Mechanical Repair & Cleaning Guide* by Curt Reeder, or *Howard W. Sams & Company Complete VCR Troubleshooting and Repair* by Joe Desposito and Kevin Garabedian. By taking good care of your VCR, you will extend its life to the point where it could gracefully age into a reliable secondary machine. You will also be able to tell if mechanical problems and breakdowns actually warrant the purchase of a new machine. Will the manufacturers get the idea when everyone stops buying junk?

A FINAL NOTE ON VCRs

VCRs are not likely to go away soon. If Super VHS takes off and/or DVD fails to deliver a recordable disk, then customers will at least have a better quality recording media. Don't hold your breath, though. Get a nice hi-fi four-head VCR for now.

LASER VIDEO DISK

Laser video disk players have remained overshadowed by the VCR as *the* way to watch prerecorded movies. However, some video aficionados have found the laser disk's quality of picture and sound worthy of their dollars.

A laser disk player uses optical disk technology: a double-sided CD-like disk the size of an old LP album. A low-power laser picks up a signal from the video disk and converts it into video images for playback on your TV, and audio for playback on your stereo or TV speakers.

The Workings of the Laser Disk. See Figure 4-4. A laser disk player uses compact optical digital disk technology. It takes an analog video and sound signal, digitizes it (turns it into ones and zeros), then records it onto a disk's spiral track. The track contains billions of microscopic pits (holes) and lands (reflective surfaces on the top of the disk). This is the digital information recorded onto a disk. For playback, a low-power laser is aimed the disk, strikes either a pit or a land, and is reflected back up to a photoelectric cell. The cell reads the digital information needed to make a picture a sound. If the laser hits a pit (hole), it relays a weak reflection back to the photocell. A land returns a strong signal. In this way, the laser disk player is fed a constant stream of the ones and zeros of binary, making it possible for you to watch exciting digitally-recorded movies.

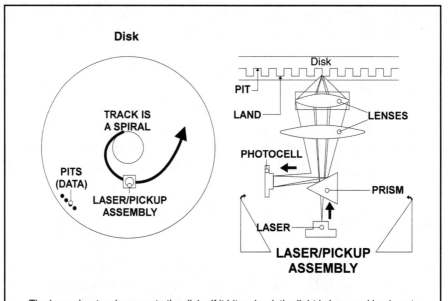

Figure 4-4. Laser disk and digital video disk mechanics.

The laser shoots a beam onto the disk. If it hits a land, the light is bounced back up to the photocell and registers as a bit. It hits a pit, it doesn't bounce back and thus produces an inverted bit. There are billions of these pits and lands along the spiral on the disk. This provides the digital information that feeds the player.

Advantages of Laser Video Disk.
- Higher picture resolution (400 lines) and exceptional quality sound: better than a VCR. The exception are the S-VHS and Hi-8 formats, for which almost no titles are available.
- There is no wear on the disk, as it comes into contact with nothing except light and the inner rotor.
- You can access any section or any frame of a movie instantly, with no noise or jitter. You can even connect a computer's RS-232 port to some players and access sections of the movie via your laptop or desktop computer.
- Sound is CD digital quality, and in some cases Dolby Digital surround sound is possible.
- On some models and formats, you can pause a frame and get a perfectly still image.
- Better special effects than with a VCR.
- As stated earlier, you can access any section of the CD right away; but this is a particularly great feature for interactive disk games and educational disks.

Disadvantages of Laser Video Disk.
- You cannot record onto video disks.
- Video stores only carry low stocks of video disk movies.
- You have to flip the disk in middle of the movie, or in some cases change disks.
- They will become obsolete as DVD takes over.

Format. Two formats exist for these laser-beam controlled saucers:

Constant Linear Velocity (CLV): A laser disk is read from the inside outward. The playing disk spins at a fairly fast speed when the laser is reading from the center. As the movie plays on, the laser moves toward the outer edge of the disk. It then slows down to lower a rpm (rotations-per-minute) in order to capture bits at a constant rate. The player automatically adjusts its rpm to account for different-sized disks. The CLV format packs more info onto a disk and can hold an entire movie on one disk. If you want *still frames* with this type of disk, you will need a top-of-the-line player.

Constant Angular Velocity (CAV): With this format, the disk's speed of rotation is constant. The scanning speed decreases as the laser pickup moves to the end of the disk. The disk can't pack as much data as a CLV unit; therefore, two double-sided disks are needed to play one movie. This format provides superior special effects capabilities, such as very stable, noise-free slow motion, and fast and slow playback in forward or reverse.

WHERE TO GET YOUR HANDS ON VIDEO DISKS

Video disks never were easy to find in stores, but don't lose hope of finding that one essential disk that you need for your collection. Sometimes you can find a mom-and-pop operation that carries a wide selection. The proprietors may prefer the format themselves and stock the disks. Frequent this kind of store. Libraries are your next best bet; don't count on any R-rated, steamy, love scene-type movies, though. Your last chance is to find people that are buying into DVD and take over their disk collection: at a reduced price, of course

Operation. The laser disk can only play back movies. Therefore, it has relatively simple controls. When you hit the *play* button, the disk spins and the laser picks up a signal much the same way as a stylus on a record player. That's it.

Sound. Because of its digital nature, a laser disk has exceptional sound. Depending on the player, you should be able to pump a Dolby Surround signal to your TV's audio system. A Dolby Digital (AC-3) track is even recorded onto some current releases. See Chapter 2 for more information on Dolby sound features.

Features. Expect to pay $400 to $1,000 for a laser disk player; although, with DVD emerging, prices are sure to plummet. There are players that ease your compatibility concerns by mixing CAV and CLV formats, audio CD formats and "CD video" formats into one device. If you are going to invest in a new machine, forsaking a DVD player, the mixed-format machine is your better choice. Other features would include:
- Double-sided play.
- Special effects.
- Digital frame storage. Great for video editing and computer-controlled digital photo manipulation.
- S-Video output.
- SCART connection.
- Video game compatibility.
- The sound could include a Dolby Surround track and possibly Dolby's newest sound tech, Dolby Digital (AC-3).

Advice. Laser disk is not likely to survive much longer. Invest in DVD if you want a quality playback device.

DIGITAL VIDEO DISK PLAYERS (DVD)

Digitally dazed high quality-video buffs are fired up and ready for this new consumer electronics release. Imagine a device that can store (and play back) an entire high-definition movie onto one audio-sized CD. One problem: video disk technology has advanced, but you still can't record onto them. Once this is possible, be assured that VCRs will become archaic video corpses rotting away in closets.

Basics. DVD is the next generation of laser disk technology; a laser disk on steroids! See Photo 4-3. It will do for home theaters what CDs did for home stereo: deliver a high-quality full-digital experience. Movies in this format are of superior video resolution and sound, compared with any other consumer media product. DVD uses a smaller laser beam, digital compression technology, and multiple layers of CD material to pack tons of digital bits and bytes onto one 5" CD.

Technology. I won't go into complicated, tongue-twisting tech words such as "cross-interleave Reed-Solomon code." Whatever it means to an engineer, it means better pictures and sound for you. With almost elflike magic, engineers are now able to pack an entire movie onto a disk the size of a common compact disk.

Like CD technology, DVD is purely digital. This means the signal coming from the disk itself is pristine, and nearly immune to any interference or wear. There is a standardized format which includes a future of two readable layers on both sides of the CD (for now, it is one layer, single-sided). This chokes the disk full of sound, picture or computer data. Also, an MPEG-2 decoder is aboard. Moving Pictures Experts Group (MPEG) is a standard that allows the video to be crunched down in size, making room for more data on the disk. A smaller red laser allows engineers to use smaller pits and lands, making it possible to store more digital data onto a given disk. All of this means that you can currently store 4.7 gigabytes (17 GB in the future) onto one disk: a whole full-length feature film.

Photo 4-3. A digital video disk player (DVD) with digital video noise reduction and a current/pulse D/A converter. Reproduced with the permission of Sony of Canada Ltd.

Advantages of DVD Players.
- Fully Digital. This means *EXCEPTIONAL* video and sound.
- Backward compatible with current CD technology. You will still be able to play your beloved Pink Floyd CDs. Compatible with DVD, audio CD, video CD and CD-R formats.
- Compact.
- Mass amounts of digital data can be stored onto a single small CD. Currently, 4.7 gigabytes are rammed into one disk. This is seven times the data in the same, neat current circular CD package; enough to hold about 2-1/2 hours of video. The storage capacity will likely increase fourfold in years to come.
- Utilizes Dolby Pro Logic sound or Dolby Digital (AC-3 sound). See Chapter 2 for more information on Dolby's digital technology.
- Unprecedented noise-reduction circuitry and laser technology.
- Digital video pictures that approach studio master tape quality.
- 500 lines of resolution.

Disadvantages of DVD Players.
- Currently, the disks are non-recordable. This will likely change when new CD recording technology is introduced in the near future. If not, it may fall victim to laser disk syndrome (recording ability was promised when the laser disk format was originally released years ago).
- Because the CD is near-master quality, movie companies are fearing copyright laws will be broken more readily. This has put a hold on many titles available for your viewing pleasure. However, as of this printing, Warner, Sony, Columbia, TriStar and MGM/UA have already released many titles on DVD, with more on the way.
- Prerecorded DVD movies bought in one country or region are not necessarily compatible with players from other areas.

Features. As of this writing, Mitsubishi, Panasonic, Pioneer, RCA, Samsung, Sony, Toshiba and Zenith have released DVD players. Some of the features are listed in the *Advantages to DVD Players* section, but others would include:
- Switches between letterbox, pan and scan, or 16 x 9 wide-screen formats.

- Parental lockout controls. This can make an R-rated movie into a G version.
- Multiple language tracks.
- Special Dolby Digital outputs, coaxial or optical.
- Component video, S-video and composite video outputs.

DVD movies are being sold for about $20 to $30, but are likely to come down in price as video stores begin carrying them.

Sound. Nothing but earfuls of digital bliss will do for this machine. This baby makes use of Dolby's newest 5.1 channel surround sound-encoded technology, AC-3. Dolby Digital is just around the corner. This is the next generation of sound sorcery that can provide the most supreme home theater experience.

The Future. Soon, with double-sided and double-layered disks, 17 gigabytes of information will be possible on one disk. In laymen terms, that is eight hours of quality digital video. This is 28 times the capacity of a standard CD. Amazing!

DVD is not the be-all and end-all of video yet. Close, though! The VCR is not likely to be kicked to the curb until DVD is recordable. When this happens, DVD will reign as the new media technology king.

WRAP UP

VCRs are likely to remain around for quite a few years because of available prerecorded movies. However, DVD is likely to soon take over laser disk technology. The video and sound quality simply have to be seen to be believed. If you own a large-screen TV, a DVD will improve your viewing pleasure a thousandfold because of the higher video resolution (which is over 100% better than a VHS).

The players are still in the $600 to $800 dollar range, so customers are not exactly buying these units in droves. As the prices come down, though, they will likely run neck-and-neck with VCRs in sales. When the prices go below $300, jump on it: the difference in DVD picture and sound is just simply astonishing, especially with a bigscreen TV.

CHAPTER 5

CAMCORDER BASICS

"Were there weddings before camcorders?" Unknown

The method of storing memories has forever changed. Electromechanical impulses now replace biomechanical ones. We are now able to take a small section of life, turn it into electronic memory tokens, then store them on a thin piece of plastic: something like the venerable photograph with a new twist. The pictures that contain our life memories are loaded with emotion through the use of motion, full 30-frames-a-second motion, not to mention the addition of sound. We are talking, of course, about video cameras. See Photo 5-1.

The modern *camcorder* saves snippets of our lives for later viewing. How does it work? Camcorders have evolved considerably over the years, with new formats, better components, and further miniaturization. Video recording cameras used to require a backbreaking entou-

Photo 5-1. A Hi8 video camcorder. Reproduced with the permission of Sony of Canada Ltd.

rage of equipment: a fair-sized video camera, a strap-on VCR, hefty battery packs, a sun-strength lighting source, and other supporting equipment. What was once a shopping cart full of valuable video gear now fits into the palm of your hand and weighs less than two pounds. What do you suppose the camcorder will like be in another 20 years?

The Essential Camcorder

A *camcorder* is a *camera* mixed with a *recorder*. In other words, the device amalgamates a VCR and video camera. Camcorders are sometimes called *palmcorders*. Because of recent advances in video photography and electronics, having a separate VCR and video camera is no longer necessary. Now a petite camcorder can be had for under $400.

Camcorders are generally classed into the format of the tape they use. VHS camcorders are still available but seldom sought because of their large size. JVC took a "Honey! I Shrunk The Video Equipment" step, creating the VHS-C format (compact VHS tape). This tape is the same width as the tape we pump into our VHS VCRs, with a shorter run-time and smaller casing. Eight millimeter, or 8 mm, is Sony's super strain of wonder tape . It packs about 2-1/2 hours of memories onto a cassette not much larger than an audio tape. High-band formats are now in vogue: they pack much more video info into an ever smaller space.

The predictable onslaught of digital has also reached camcorders with resounding results. More on this format later.

The video camera part of the camcorder is made up of many subsystems: the lens, the CCD or imaging device, the viewfinder or LCD display, sound components and the support electronics. Let's dissect the innards of the camcorder to better realize its workings.

FORMATS

Camcorders are generally classed according to the type of tape used and the format used. For example, an *8-mm camcorder* uses 8 mm tapes for recording. Formats have different resolutions, recording quality, sound quality and different recording times.

VHS. The immortal VHS format was created by JVC. Problem is, the physical size of the tape does not make for a compact unit, and the video resolution is terrible. Camcorders that use this 1/2" format are still available, albeit hard to locate and unpopular. One advantage to owning one is that you can take the VHS tape out of the camcorder and plug it directly into your VCR. Another advantage is that it provides a long recording time; 2 to 6 hours is typical. Last but not least, there is the fact, "The heavier the camera, the more steady you can hold it."

VHS-C, or Compact VHS. This format was JVC's brainchild as well. It is actually the same width tape as a standard VHS tape, 1/2" or approximately 13 mm. However, there is not as much tape spooled onto the cassette. The physical dimensions are about one third the size of a standard VHS cassette, so it allows for the engineering of very small camcorders. The trade-off is, of course, shorter recording times: 30 to 45 minutes at fast speed (SP), 60 to 90 minutes at slow speed (EP). However, there is a new breed of tape that extends this to 40 minutes fast and two hours slow. Others are on the way. More on these later.

> **NOTE:** A tape can be recorded at different speeds. SP (standard play) is a fast-recording speed that uses a lot of tape to give high-quality recordings. EP (extended play) is a slow-recording speed that uses less tape, but the quality of the images is somewhat low. LP hangs in the middle of SP and EP.

There is an adapter available that looks like a standard VHS cassette with a huge hole in it. You place the VHS-C cassette into it and shove the unit into your VHS VCR for playback. This makes playing the small VHS-C tape more convenient than using an 8-mm that you cannot plug into a VHS VCR.

Video-8 or 8 mm. After the death of BetaMax, Sony went with this new format. See Photo 5-2. It is an 8 millimeter-wide tape wound into a case the size of an audio cassette. Unfortunately, you cannot play these tapes back on a VHS or S-VHS VCR, even with an adapter. The system was designed to be a compact dynamite video machine. Video-8 offers better quality video and sound than VHS-C (in a smaller package). Another advantage to 8 mm is that it will record up to 90 minutes on SP and 180 minutes on LP. Larger capacity tapes are continually being released. Remember, using LP will degrade picture and sound quality.

High-Band Camcorders. These camcorders pack more information into the same amount of space as their brothers, the VHS-C and Video-8. They also provide better color separation and brightness with lower

Photo 5-2. An 8 mm camcorder, with a 270 degree rotating LCD screen. Reproduced with the permission of Sony of Canada Ltd.

signal-to-noise ratios. The quality is so good that high-bands almost match the $100,000 units broadcasters use. In fact, some TV news crews tote these gems around instead of a broadcast camera, for fast-breaking stories.

S-VHS-C, Super VHS Compact: If you wanted a high-resolution camcorder, this may be the answer. The tape is similar to the VHS-C, but records more picture information. The S-VHS-C reproduces 400 lines of resolution for stunning shots. The problem is, you can't play the tape back with a conventional VCR or adapter. However, it will work if you have a S-VHS VCR, adapter and high-resolution TV. The question is, do you want to spend big money on a whole new system? If you use professional-quality video and editing, go for it. Otherwise, save the few thousand dollars for more important items.

Hi8: Review Photo 5-1. This high-quality format typically offers 40% greater resolution than standard 8 mm. Very popular! Hi8 gives near-professional broadcasting-quality recording. By utilizing special metal-evaporated tape (Metal-E), Hi8 pumps up the video and sound quality past S-VHS-C. Just as the S-VHS-C is incompatible with other systems, so is the Hi8. Investing in tons of high-resolution equipment may be in your future if you choose this format.

A CAMCORDER'S VCR COMPONENTS

About the only differences between your VCR's workings and those of your camcorders are as follows:
- Compact camcorders tend to use a smaller video head. However, they wrap the tape around it further and increase the rotation speed. There is a trend to move back to full-sized heads for a more accurate picture when transferring over to standard VHS tape.
- Camcorders use four video recording heads and one flying erase head. The four recording heads are used for better video effects. The flying erase head is like a surgeon's scalpel. Instead of a chain saw-type of VCR erase head that takes away everything on a tape in one whack, the flying erase head takes out much smaller sections.

CCDs, MOS and CMOS

A camcorder does not use photographic film. Instead it *sees* by utilizing an electronic imaging device called a CCD (charged-coupled device). Light makes its journey through a series of lenses, then strikes the CCD, creating an electrical impulse. This is then sent to the VCR within the camcorder or to other circuits.

Past. In the past, video cameras used vacuum tube-imaging devices. These bulky, unreliable, power robbing units needed a Niagara Falls-sized turbine generator to run, and used more lighting wattage than the Las Vegas Strip. Running the show was a 4" long tube that contained parts very similar to a TV's picture tube.

Present. Solid-state technology has taken over in the camcorder imaging process. A CCD sits in almost every camcorder on the market today. This is a light-sensitive integrated circuit measuring about 1/3" square. Another solid-state solution to imaging is MOS, or metal-oxide semiconductor. MOS functions are basically the same as for a CCD. Some manufacturers prefer one over another.

The semiconductor (CCD or MOS sensor) is divided into hundreds of thousands of segments called *cells or pixels*. Each cell reacts to the light aimed at it by the lens system, then converts the light into the appropriate electrical information. Each pixel's signal represents a color and its intensity. One semiconductor chip typically has 250,000 pixels in older units. Current ones average 270,000 pixels. Some even go up to 410,000 to 570,000. There are a few companies that are doing work with CMOS (complementary metal oxide semiconductor) sensors that are around 800,000 pixels. Look for these in the near future as they are very cost-effective.

What do these numbers mean to you? Probably not much. A 250,000-pixel unit will reproduce a typical VHS picture. The new Hi-8 and S-VHS-C formats make use of 410,000 pixels, converting your home movies into pictures with superior clarity and sharpness. The problem is, you need a TV that will display over 400 lines of resolution, and a VCR can play them.

So, on what should you spend your money as far as imaging resolution goes? If you want a camcorder to record the kids' activities and your vacations, save yourself the extra expense and go for a standard unit that uses 250,000 pixels. If you want near-broadcast quality, then go for the Hi-8 or S-VHS-C, which use 410,000 or more pixels. However, expect huge expenses for the support equipment.

Lux. Light entering the camcorder is measured in *lux*. This is the light's *intensity*. If you place a candle one foot in front of you, it will produce 10.76 lux (one footcandle). CCDs are rated according to lux. So, if you buy a CCD that reads 2 lux, then a candle can provide about 5 times the light required for viewable movies: impressive compared to older 2000 lux imaging devices. So, if you are looking for a camera for filming low-light scenes, pay attention to this number.

LENSES

Light is an enchanting form of energy. A lens allows you and your camera to play many illusionary tricks with light. You can bring an object's image closer and focus it to perfect clarity. You can widen a field of vision, bringing in all of the nuances of the panoramic scenery, or simply cap it off and call it a day. What kind of lenses are standard on camcorders these days?

Fixed Focus. Not too many fixed focus lenses are left out there. To slice costs, some manufacturers cut auto focus features. They are similar to a simplified, point-and-shoot film camera.. Do your legs and sanity a favor and forget buying one of these units. They are very limited to the depth of the video pictures you will be able to take.

Variable Focus, Auto Focus. Most manufacturers produce units which can focus on a subject either with a manual adjustment or automatically. An infrared beam is emitted from the camcorder, bounces off a subject, then is reflected back at the camera. It helps the camera determine the subject's distance and automatically dials in the correct focus setting. A manual override is a must as sometimes the infrared beam will lock onto the wrong subject, throwing off the focus.

Zoom. Close-up, far off, or close-up far off, you have to love this overworked feature. It magnifies the image you are looking at by up to 40 times the original picture size. A recent zoom technical advancement is *digital zoom*. It will simulate a picture up to 200 times the size of the original.

The digital zoom feature is measured in X; 12X is twelve times the size of the original zoomed-out picture. Look for a camcorder with a minimum of 12X mechanical zoom, and if possible get one with digital zoom. The larger the X, usually the more costly the camcorder.

Wide Angle. This type of lens expands your view of a picture's width. It is useful if you have to be close to a subject but need a wide shot. Most camcorders have this feature built into the zoom lens.

Macro. This feature lets you get the camcorder extremely close to a subject and still be able to focus. One example of this lens' use would be when photographing a rose's delicate interior. Another use is to make a videotape of a photograph for photo montages. Most camcorders have this feature built in.

Detachable Lenses. An extreme rarity these days; if you want to purchase a camcorder with a removable lens system, stay away from the brands which make you purchase their "special lenses." Costs can run over $150 per lens. Instead, look for a camcorder that utilizes a C-mount. This will let you interchange common (and cheap) 35 mm lenses and filters.

VIEWFINDER

It's nice that the camcorder can take in what it has to, but what about someone else being able to see what's going on? The *viewfinder* is the device through which you can view the scene you are videotaping: your window to the recordable world!

Electronic Viewfinders. Most camcorders contain a small LCD screen (liquid crystal display) inside the viewfinder. It is a miniature TV that shows you what the camera is picking up. The good thing about this function is that you can run a tape back a bit and watch what you just recorded through the viewfinder. Depending on the age and cost of the camera, the LCD is either black-and-white or color. Do your eyes a favor and opt for a color unit.

Camcorders come with battery and tape indicators that feed their signals to the LCD. This puts the indicators and titling messages in your immediate line of sight.

HINT: When using an optical viewfinder, back away just slightly from what you are recording. You may take in a bit of extra scenery, but it is better than cutting off someone's head just because the viewfinder was not correct.

The new rage is miniature screens which either fold out or are attached to the camcorder. See Photo 5-3. They are similar to laptop computer screens but range from 2" to 5" in width. Here are a few advantages to these units:

- Other people can see what you are recording or have recorded without hooking the camcorder to a TV. Disadvantages are the same as previously mentioned.
- You can shoot without having the camcorder permanently stuck to your eyeball.
- You can swivel the unit 180 degrees and become part of the scene via remote control.
- Editing will require one less TV or monitor.

Photo 5-3. A video camcorder with an attached LCD viewing screen. Reproduced with the permission of Sharp Electronics Corporation.

These units are in black-and-white and color flavors. Leave this option decision up to your pocketbook. The color LCDs are expensive.

Optical Viewfinder. If you have ever owned a cheap 35-mm film camera, then you are familiar with optical viewfinders. The "hole" you look into is separate from any other lens in the camera: it's usually just a through-hole system. This is fine if you don't need to know precisely what is in the frame while recording. If you want to cut a few dollars from your camcorder budget, go for this option.

Through-the-Lens Viewfinder, TTL. This type of system is similar to the optical viewfinder. It takes the image (light) coming in from the main lens, splits part of it up to the viewfinder and sends the rest of the image to your eyes. In other words, 80% of the light goes to the CCD, and the remaining 20% is routed to the viewfinder with mirrors.

HINT: Keep in mind that the through-the-lens viewfinder feeds 20% of the light to your eyes. This means that you will need to take a little more care in filming low-light scenes; you have to feed the CCD a little extra light than you would for an LCD-viewfinder camcorder.

The disadvantage to the last two viewfinders is that you have no way to preview your shots while still in the field.

FEATURES

Here are a few standard features you should demand in your camcorder:

Image Stabilization. Jiggle! Jiggle! Few things are worse than watching an unsteady scene. It makes your home movies shout, "Amateur!"

Chapter 5: Camcorder Basics

HINT: If you want a mechanically steadier picture, go for a larger camera or at the very least a tripod.

I know it is difficult to keep a steady hand, so companies have come up with a partial solution: image stabilization circuits. They electronically steady the results of a shaking hand.

Remote Control. Great gadget for making yourself another video-recorded victim. It lets you get in front of your own camera. Also, if you are going to be editing, a remote control is a must! Look for one with easy-to-use, well-placed controls.

AA Battery Compatible. Some palmcorders use a battery pack, but some also have a built-in compartment for AA batteries. This is great if you are running low on the battery pack and don't have an extra or time to recharge. Just pop into a 7-11 store and appropriate a few AA batteries.

See Chapter 9 for further camcorder features.

SOUND

People rarely consider audio options when purchasing a camcorder. The audio is, in fact, half the experience. Imagine videotaping your baby's first words with it. Imagine a rocket-fueled drag race with the blasting thunder of the engines in full stereo. Only a few sound options are available to you, but they will make a world of difference in your videos.

ALC. All newer units come with an *automatic level control*, or ALC. Said out loud, this function even sounds like the howling wind or a kid yelling into the microphone. It also jacks up soft sounds a bit, like a cricket chirping or a whispering voice.

Normal Mono. Most cameras use a single source to record sound. Usually it is a microphone mounted on the top of the camcorder. Most cameras also come with a jack so you can plug in an external microphone. This will move the microphone away from the racket of the camcorder's clicketty-clacking innards. If you are just going to be recording speaking voices, these camcorders are just fine.

Hi-Fi Stereo. This feature is usually found on Hi8's or S-VHS-C units. It's great for reproducing music at concerts. Otherwise, use the extra cash to buy more blank tapes.

DIGITAL CAMCORDERS

The digital invasion has taken over camcorders as well. Several manufacturers such as JVC, Panasonic, Sony and Sharp now offer high-end camcorders that use compressed digital video and sound. For the time being, this is a diehard videophile product. A digital unit looks almost the same as a low-end VHS-C or 8 mm unit, but be assured that the price tags will be different. Digital units are currently going for around $2,500 to $5,000, but are likely to come down soon. See Photo 5-4.

THE DIGITAL DIFFERENCE!

Here is a rundown of the features that are proprietary to digital camcorders:

Format. These video binary number-eaters are fueled by a new type of tiny tape called *mini-DV*. This is a new video industry tape standard which comes in 30-minute or 60-minute lengths. Some of these tapes offer a 4K memory chip that stores information about the tape's length, the recording date, position of the first scene, and a still picture table of

Photo 5-4. A digital video camcorder. Reproduced with the permission of Sony of Canada Ltd.

contents. The ultra-compact size is truly amazing: try half that of a standard 8-mm cassette or 1/12 that of a standard VHS cassette. *A MERE MATCHBOOK!*

CCDs. Most digital camcorders use three internal CCDs: one for each primary color's beam of light (red, green and blue). Some companies offer a cheaper version with only one high-resolution CCD.

Video Compression. Compression is a way to pack more video information into a small amount of space. By translating the video signal coming into the camera to a series of numbers, the camcorder's processor is able to correct errors, then crunch the information down to one-fifth its original size. From there the information is stored onto a teeny-tiny cassette tape.

Time Codes. This feature brings professional video editing capabilities into your hands at home.

Sound. Audio circuits didn't escape the digital revolution either. Digital camcorders use *pulse code modulation stereo* (PCM). This is a way to record CD or DAT quality sound. Some companies have a dual-mode PCM sound system, which allows you to record either two channels of super high-quality sound or four channels of reduced-quality sound.

16:9 Modes. To create a cinematic feeling, manufacturers added the ability to record in a widescreen ratio.

Video Resolution. Digital camcorders display 500 lines of resolution, but you need a high-resolution TV to view the picture. This is a 25% clarity improvement over S-VHS and Hi-8 formats.

RS-232 Port. Some of the camcorders are able to download still images directly to your computer for digital manipulation and storage. Beware, though: most companies currently are not offering this feature. Your only option would be to purchase an expensive video capture board to import still pictures. Look for changing standards in the future.

THE DIGITAL CAMERA REVOLUTION

Old photographic film technology is being replaced by digital technology. Where there was once film, cameras and expensive developing costs, there is now a digital camera which stores pictures as bytes for use with computers.

A digital camera is similar to a camcorder except it stores single images onto a memory chip, flash memory card or portable hard drive. It uses a CCD similar to a camcorders to record images which it then compresses to as few bytes as possible and stores. Once you have taken as many pictures as the camera's memory can store, you can download the digital images into your computer via a special cable. From there you can manipulate the photos, group them, and output them to a TV or Web page, or print them out with a color inkjet printer or color laser printer. If you use special glossy photo paper, the printer will spit out a near-photographic quality hard copy. Quite stunning!

Digital cameras are being used as a quick-and-dirty way to grab shots for later digital manipulation. Real estate offices use them for listings, families use them for Web page picture postings, and companies use them for inventory. In fact, most newspapers use high-end digital cameras so they can download pictures from anywhere in the world for immediate printing.

Here are a few facts and figures to help you find the digital camera that's right for you:

Resolution. The first figure to look at while considering a digital camera purchase is resolution. This is the number of pixels (dots) that make up one image. Currently, a digital camera has quite a low resolution, usually starting at 320x240. An average resolution is 640x480. This number will surely go up with time. Look for the highest resolution your budget allows. For simple vacation shots and such, look for a 640x480 model.

Storage. Most of the new, simple digital cameras have built-in storage of about 20 to 50 images, depending on the resolution mode you use (the lower the resolution, the more photos you can store at a time). Some units have disks you can swap in the field, and some let you attach a portable hard drive, letting you store 3,000 images plus audio snippets.

Display. Your options for a viewfinder are identical to those of a camcorder. There is a simple optical through-hole viewer, or an LCD display which lets you view all of the pictures that have been stored. An LCD unit is generally 1/3 to 1/2 the cost for an optical model. Do yourself a favor and get at least a low-quality LCD model so you know what images you have taken already.

ADVANTAGES AND DISADVANTAGES OF DIGITAL CAMERAS

Digital camera technology is not quite perfected yet. The resolution has to increase manyfold before it replaces the venerable 35 mm film camera. Here is a list of other advantages and disadvantages in this accelerating market:

Advantages. Zero turnaround time for processing. No scanners are needed to digitize photos. The small digital camera is handy and slightly more portable than a camcorder. It is easy to download the images into your computer.

Disadvantages. The resolution is unacceptable for professional photographs. The storage capacity is currently limited. Even the cheaper versions are very expensive, starting at $400. Professional models for newspaper use are currently running $15,000 to $75,000. Memory cards are even more expensive, running hundreds of dollars for only a few megs.

DIGITAL CAMERA ALTERNATIVES

A camcorder and a capture box such as the Snappy by Play, Incorporated, can produce just about the same results as a digital camera. The only downside to this is that using this combination is a long, complex procedure.

Some manufacturers are now integrating digital cameras into their high-priced digital camcorders. When the cost comes down, this is likely to become a VERY HOT product: video and stills in one package! The digital future is just one "click" away.

HINT: While downloading your photos to the computer, hook up an adapter to the camera to save the batteries.

Advantages. If money is no object, then what advantages are there to purchasing a digital camcorder over a Hi-8 or S-VHS-C?

Weight: There is a slight weight savings in a few of these units. JVC and Sharp seem to have won the battle of the bulge with units as light as one pound.

Clarity: The resolution of these units is almost on par with top-level broadcasting equipment.

Ultra-low Noise: Video noise levels are 200 to 300 percent lower than with current analog home video equipment.

Faithful Color Reproduction: With three CCDs, a true color reproduction is possible. Current camcorders sometimes drop colors while recording; not so with digital.

The Future of Digital. Currently these units still only have RCA-type ports (A/V), and in some cases S-video ports. This means we are one step from being able to hook a pure digital line to a computer for a professional quality home editing system. JVC and Sony are making inroads into these features. When this happens, digital camcorders will take over the market. In conjunction with your computer, they will become an economical all-in-one package for making your memories last forever.

CAMCORDER ACCESSORIES

When purchasing a camcorder, it is best to pick up a few extras. Don't go overboard yet, but make sure you won't get stuck at a bad moment without a vital piece of equipment.

Cases. A case can be one of the most important accessories: somewhere to safely put everything. Look for a case that offers good shock protection for the delicate electronics. Another hint is to buy a case that doesn't look like a camera case and has no imprinted logos. This is a valuable theft deterrent.

Batteries. Ugh! This is a hard one. Rechargeable battery packs can be pricey, upwards of $120. Is it worth it to have an extra pack? If you have a camera that supports AAs, you may want to go to K-Mart and simply buy a few rechargeable batteries and a charge pack. They are about $2 to $5 a shot, a big savings over $120.

Lighting. Some cameras have a mountable light. This is great for low-light situations, but lights are power robbers. Carry extra batteries or buy a rechargeable with a higher amp rating.

Tripods. These are a must for steady images and remote-controlled recording. Prices range from $35 to thousands of dollars. Look for one which moves very fluidly. This will prevent jerky shots while panning up-down/left-right.

SUMMARY

The technology of video cameras is continually moving toward cheaper, smaller models. Now with digital, a camcorder will form a more melded package with your computer for video editing. With a Hi-8 and a computer, it is now possible for anyone to produce quality productions. This is verified by a recent interview with George Lucas, who himself got started with such small low-budget features as *THX-1138*. So, with some smart camcorder shopping decisions and a good movie idea, who knows what you can accomplish?

CHAPTER 6

VIDEO & AUDIO EDITING

*"The difference between an amateur and a professional is that an amateur shows you **ALL** his pictures."*   Unknown

Many Hollywood movie directors started with a home movie camera or camcorder. Students, adults, and practically everyone with an entrepreneurial spirit would like to follow in Steven Spielberg's footsteps. Learning the trade becomes easier as the technology becomes more accessible and cheaper. Now it is possible for anyone with a camcorder, a VCR and (sometimes) a personal computer to make dazzling, panoramic, heart-stopping home movies. George Lucas himself uses a Hi8 to do preliminary scenes. You can now easily film multiple scenes, edit them, add sounds, fix video boo-boos, paste in fades, create video effects, and (the fun part) add special effects and titles. Imagine filming the next *Star Wars* sequel with a $500 camcorder! Will your video productions ever be on par with Spielberg? Maybe, maybe not: but everyone has to start somewhere.

EDITING PRINCIPLES

Here are some basic concepts and information about video editing:

How the Big Boys Do It. Most professionals in broadcasting and the film industry have made a transition to computers to edit films. They take the raw video footage, digitize it (turn it into a format the computer will understand), then save it onto massive hard drives in a computer. From there they add titles, effects, fades, additional soundtracks, and so on. After editing is complete, the newly packaged video is pumped back out onto videotape or film. It is predicted that within two to five years, the ENTIRE process of moviemaking with be through digital technology (digital cameras, computers, playback devices, etc.).

Reality. Unfortunately, the digital systems can cost hundreds of thousands of dollars. Therefore, the average video consumer is stuck with age-old video editing techniques. However, two VCRs and some simple processing equipment will suffice. One VCR and a camcorder will work also.

As of this writing, however, there is a move by companies such as Avid Technologies and Apple Computers to bring pure digital editing into the nonprofessional market. This means for around $500 to $700, you can purchase a video capture board and software to edit home videos right on your PC. As of this writing, the system is only available for Apple's Power Macintosh and the Performa series of computers. Look for these technologies to come to all PCs in the near future.

TYPES OF EDITING

As we said, there are several techniques available to the ordinary public:

In-Camcorder Editing. This is nearly impossible to accomplish. In-camcorder editing is shooting the scenes in the exact order in which your movie will appear. This means erasing any bloopers and blunders each time they occur by backing up the camcorder's tape position and filming the botched scene again. If you use a storyboard, this method may work for you.

Filming to Edit. This is a more realistic technique. Film everything; then after you have all the footage, including mistakes, edit the film with the two-VCR method or with a computer.

Post-Production Editing. With this method, when all of your tapes are filled full of footage, then it is time to edit. This is also called the post-production stage. It's your chance to fix goofs, add pizzazz and have fun.

METHODS OF EDITING

Many methods of editing exist these days. Gone are the days of physically cutting and pasting videotape. Now barrages of buttons and millions of multiple-format tapes are the fare.

You can selectively copy good scenes from one tape to another in an order that creates a story. This can be done with a camcorder and a VCR. An alternate system would be more complicated: computer-controlled video processors and digital video systems.

Manual Editing (Two-VCR Method). This is the cheapest possible option available to budget-conscious videophiles. All you need is a patient, well-worked finger, a camcorder and one VCR. If the camcorder you used for filming was a rental and has already been returned, you may use a cheaper VCR or video cassette player in its place. With the raw footage poked into the camcorder (or a secondary VCR), here's what you do:

1. View all of the footage a few times and create a storyboard. List the scenes you want to include as well as dump. See Figure 6-1.
2. Using the camcorder or VCR's tape counter, note the number where each scene begins and ends. Write this on the storyboard:
   Example:
   Scene 1, At beach with kids. Start 000147. End 000741.
   Scene 2, Sunset on the mountain. Start 003000. End 003100.
   Scene 3, Close up of kids. Start 000741. End 000981.
   Scene 4, On the way home. Start 001000. End 001600., etc.
3. Hook up the camcorder's *OUT* audio/video jacks to the second recording VCR's *IN* A/V jacks. Some camcorders have special connectors; check your manual. Make sure to switch the VCR's signal source to the *IN* jacks or you may record another channel without realizing it.
4. Insert a blank tape into the VCR for recording your newly edited home movie.
5. Advance the footage tape to just before the numbers you wrote previously on the storyboard.
6. Hit the appropriate *RECORD* buttons on the recording VCR. Quickly hit *PAUSE*.

Figure 6-1. Making a storyboard.

7. Hit *PLAY* on the camcorder and get ready to unpause the VCR.
8. When the scene is where you want it, release the *PAUSE* button and let the tape record onto the VCR. Hit *PAUSE* on both machines when the scene is done.
9. Continue this process for each scene until your home movie is complete.
10. Watch and enjoy the movie!

This method is fine for low-quality editing, but it takes a bit of practice. Have patience.

Using Editing Controllers. A cheap useful addition to your miniature movie studio is one of these controllers. See Photo 6-1. It takes the place of your fingers having to fumble with buttons. A controller hooks to your camcorder and recording VCR. It then *controls* the buttons remotely and automatically. Some editing controllers use cables and others use the remote control's wireless circuits (no wires needed). You can watch the raw footage and make marks on the scenes you want to keep, the ones you don't want to keep, and in what order you want them to appear. You then let the controller automatically *compile* the video for you. In other words, the device advances the tape to where the scenes are and records them automatically: an *autopilot* for video editing! A terrific first upgrade.

Some companies offer these units as a stand-alone box or as a peripheral to add to your computer. Editing controllers typically run from $99 to $1000. Gold Disk, Inc., offers a computer-controlled unit with cables and software for $99. An upgraded version is $199. The Thumbs Up Thumbs Down unit from Videonics is a stand-alone unit that sells for $100 or so.

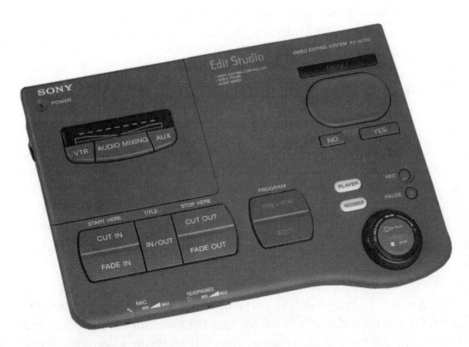

Photo 6-1. An 8-mm video editing system. Reproduced with the permission of Sony of Canada Ltd.

Chapter 6: Video & Audio Editing

The Edit-Control Feature on Your Camcorder or VCR. A feature called Control-L , Control-M, Control-S, Control-T or LANC (depending on the manufacturer) allows for two-way communication between the VCR, camcorder and editing controller. You can remotely control the VCR and camcorder for editing purposes. If you are starting a video package from scratch, purchase these options on your VCR and camcorder now and save a few hundred dollars for additional editing equipment. See your manufacturer manuals for operational procedures.

Digital Editing. Video editing now has a consumer digital solution. For the last few years, digital editing was only in the realm of broadcasters; but with advances in computer storage and processing power, consumers are now able to afford their own digital editing packages. These include a video capture board and software. The caption board digitizes footage then stores it onto your computer's hard drive. With a few programs, and you have just about every editing and special effects feature available to you. Mix the scenes, edit, add characters and sound then, send it out to a VCR. You can also watch it right off your HD.

The cost of a typical digital editing package is in the $500 to $5,000 range right now. However, most video graphic card manufacturers are now building these units into their high-end video boards. It's a great opportunity to save on editing equipment.

LINEAR vs. NONLINEAR EDITING

Refer to Figure 6-2.

Linear Editing. This means to edit a videotape by accessing one scene at a time on one tape, then transferring it to a master recording tape. In other words, editing by using two VCRs or a camcorder and a VCR.

Nonlinear Editing. Editing by using a computer and its hard drives.

Pros: Linear editing offers the ability to add equipment one piece at a time, and it's a cheap solution for simple home movies with a few effects. Nonlinear editing allows for fast compilation of the video and better ability to control the process. Special effects are also much easier

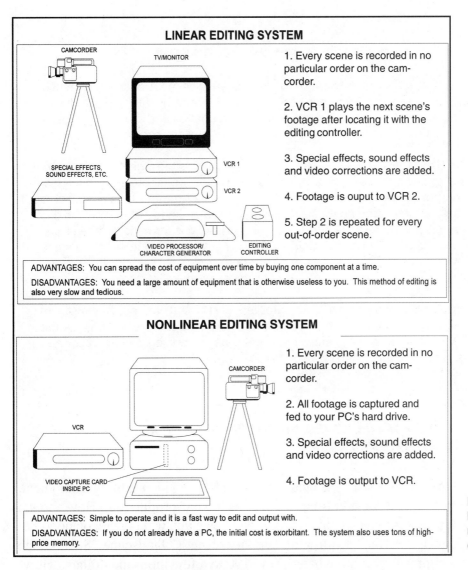

Figure 6-2. Linear vs. nonlinear editing.

in the digital realm of nonlinear editing; no need to add more expensive, bulky equipment after the initial purchase of the video card and software.

Cons: Linear editing is slow and tedious, and special effects are limited to the equipment on hand. It also requires many more pieces of equipment to achieve the same results as one computer and a camcorder. Nonlinear editing has a high start-up cost and uses mass amounts of memory: both RAM and hard drive space.

DATA COMPRESSION AND MPEG EXPLAINED

Digital video signals can be compressed considerably. Since most video equipment is now digitally-based and runs with microprocessors, it opens up many digital tricks. Compression is a way of taking a large amount of information and crunching it into a much smaller size so you can transmit it faster and use less storage space.

Motion Picture Expert Group (MPEG) is a standard that allows the digital compression, transmission and decompression of a video signal. MPEG works by taking each picture frame and breaking it up into sections. When the next frame is displayed, some sections are exactly the same as the last frame, such as an image of the sky. In this case, the information from each section is not needed from one frame to another; it just reloads the same information as the last frame. This saves having to send the same information over and over. The only picture information needed are sections that change from frame to frame, such as mouths moving or a spaceship rocketing across the screen.

MPEG is used in digital camcorders, DBS satellites and DTV. It can reduce the size of the video tape in a camcorder to the size of a matchbox. It packs 6 to 10 times the amount of TV channels into the same bandwidth as an analog signal. It makes videophones possible using standard phone lines.

TITLE MAKERS, VIDEO/SOUND EFFECT MIXERS & VIDEO PROCESSORS

One step up the cost ladder is *stand-alone* video editing equipment. With it you can add titles and credits to your production. How about a few professional quality dissolves? Maybe you want to videotape someone against the backdrop of "another planet's" landscape. Stand-alone equipment runs anywhere from $100 to a few thousand dollars. However, if you are doing semiprofessional or professional quality video, it may be the most logical choice.

Title Makers or Character Generators. *Cost: $300 and up.* These units let you add high-resolution text to your productions. Look for a model with multiple fonts and font effects, such as shadows, multiple languages (if needed), and NTSC support for North American and Japanese video editing. (Remember, a person in England cannot view a videotape created on an American VCR.)

Video Mixers. *Cost: $500 and up.* Hundreds of video effects are possible with these units. Look for one that can handle wipes, cuts, dissolves, black-and-white pictures, negative pictures, paint, flip functions and audio mix. Another great feature is one called *chroma key* or *blue screen*. It lets you replace a single-colored background with any image you want, just like a TV weather forecaster shot.

Make sure the video mixer unit has multiple inputs so you can hook up more than one source. Some of these units even have preview screens that you can use as additional preview monitors.

Video Processors. *Cost: $1000 and up.* These units *fix* or alter the video palette (colors), enhance video images and analyze video signals. With a video processor, you can pull apart any section of the signal and do a patch-job. So, if you have video footage that is absolutely terrible, this unit can work miracles. An example would be repairing low light or that terrible yellow/green tint from fluorescent lights. It can also take a weak signal and amplify it. This is great if you are recording a tape from one generation to another and losing quality as a result. Analog previously prevailed; problem was, you needed a mile-high pile of units just to make a few simple video corrections. A digital video processor is an all-in-one package. Look for a unit that has a video processor, analyzer, color processor and a video stabilizer built in.

Sound Effects Mixers. *Cost: $300 and up.* Add POPS, ZINGS and BANGS to your videos. Look for a unit that lets you alter and add soundtracks, and has multiple audio inputs for hooking up various stereo signals sources.

Videonics, Inc., manufactures just about anything having to do with digital editing, including mixers, processors, etc. Check out their Website at http://www.videonics.com for tons of information on editing.

PUTTING A PACKAGE TOGETHER

You can piecemeal a package together over time, or you can feast like crazy by buying every top-of-the-line item available. For simplicity's sake, let's look at one-at-a-time packages:

Camcorder. The camcorder is the obvious beginning of a home movie system. Look for as many features and effects as you can afford right now. Recording features can replace thousand-dollar editing items you would need later, such as fades, titles, Control-L, remote control, flying erase heads, etc.

VCR. This choice will depend on the quality of the video you want to create. Do you really need to spend $1,500 on a Super-VHS package, or will a $200 bare-bones K-Mart special do? You may consider purchasing the same brand name as your camcorder because some features are dependent upon the VCR/camcorder symbiosis. Some VCRs have a features package that is specifically directed at home movie editors. Be prepared to pay double the price of a normal VCR.

EXPANDING YOUR LINEAR SYSTEM

Start by adding a simple editing controller such as Videonics Thumbs Up 2000. It will give you more command over your VCR and camcorder, and will allow some functions to be set on semiautomatic.

Next you will want to invest in a video mixer to create video effects: wipes, cuts, chroma key, etc. As money allows, add title makers, character generators and sound effects mixers. If you are having problems with low-quality video footage, look into a decent video processor.

If you will be doing professional videos, I suggest investing in a special editing VCR. An S-VHS editing VCR will set you back about $2,000. A digital one will kill the video budget for the rest of your life at about $12,000 to $15,000.

EXPANDING YOUR NONLINEAR SYSTEM

Most people who own camcorders own home computers as well, PC or Macintosh. By purchasing a video capture card and/or software, you will be able to edit like the pros without having to go through the constant play, pause and record routine. In the future, most computer manufacturers will most likely build these capture boards right into the motherboard. Until then, the initial costs are somewhat high.

System Requirements. Requirements will vary for each manufacturer's system, but this is a middle-of-the-road guide:

Minimum: Pentium Class PC, 16 megs RAM, 500 megs of free hard drive space, Windows '95 operating system.

Preferred: Pentium 133 or faster, 32 megs RAM, 2 gig SCSI hard drive space, 24-bit accelerated video card, Windows NT 3.51 or higher.

For Macs add OS 7.5x or above, PCI-based Power Macintosh.

Disk space required for editing (approximate): 5 min. = 225MB, 10 min. = 450MB, 15 min. = 675MB, 20 min. = 900MB, 25 min. = 1.1GB, 30 min. = 1.35GB.

Companies to Contact. Avid Technology, Inc. (Avid Cinema), http://www.avidcinema.com; Digital Processing Systems, Inc. (DPS), http://www.dps.com; Miro Computer Products AG, http://www.miro.com; NewTek, Inc. (Video Toaster Software), http://www.newtek.com.

Costs. $500 to $4000 for the capture board and software only.

STEPS TO LINEAR EDITING

Do you have a stack of videotapes that no one would want to watch? The solution is to edit. Besides cutting and pasting the good and the bad scenes, editing can also be used to add effects and titling, and to correct improper or poor footage. Here's how it's done:

1. Set up your equipment. Put the footage tape in the proper piece of equipment and the blank tape into the recording VCR.
2. Storyboard your movie. Cut and paste the footage in the order you want onto the new tape.
3. Add titling. (Optional.)
4. Add special effects such as wipes and fades. (Optional.) Don't overdo these, or it can turn into a true show of amateurism.
5. Change the soundtrack or add music and sound effects (splat!). (Optional.)
6. Chroma key additions. (Optional.)
7. Duplicate and distribute. (Optional.)

STEPS TO NONLINEAR EDITING

1. Transfer your video footage onto your computer hard drive by digitizing it through a video capture board.
2. Open up the editing software and hack away!
3. Output the video to a VCR loaded with a blank tape.

A FEW TIPS ON MAKING HOME MOVIES

Here are a couple of tips and hints on how to make your home movies a HIT! First of all, as you now know, editing means to take what you've shot, cut and paste sections in a logical order and add special effects. However, the whole process really starts in conveying a story to the viewer. What story do you want to tell? If the video consists of vacation shots, was there a central theme to your travels? Adhere to this theme when editing your video.

Also, make sure to take detailed notes as you film so that you know which scenes are priceless memories and which need to fall victim to an erasing head. Tell a story that you will be proud to show friends and family members. Don't be afraid to get artsy and let loose the movie director in you. Creativity in home movies is an asset, and can make a huge difference between a movie you want to watch again and again, and one that ends up in the used tape pile.

WRAP UP

Fifty-five million camcorders are taping away throughout the world. In order to make polished productions out of all that footage, you need to know how to edit. By taking care and adding an effect here and a soundtrack there, you will be able to captivate audiences instead of boring them beyond belief.

Start light and add to your little home studio a piece at a time. Try to invest in a nonlinear editing package, as it will be cheaper in the long run and more compatible with most future video technologies. Action!

CHAPTER 7

VIDEO GAME & INTERNET CONSOLES

"Do you want to play a game?" WOPER in *Wargames*

Even back in prehistoric days, humans created activities and games to occupy their time and enjoy themselves. Whether it was racing Flintstone-mobiles or batting a bedrock with a bone bat, we kept busy. Then something strange happened: the Romans came along and taught us that being a spectator was a sport of its own.

Now, in the electronic age, we have realized new ways to play games without having to traipse across town to sit in a stadium. You can now sit in the privacy of your own home with a TV and a video game console, and take control of your favorite heroes and battle the evil minions. If that doesn't sound appealing, how about a relaxing game of football, in which you usually end up screaming at the referee like a real football player?

The video game industry has come a long way since the introduction of the Atari and later the original Nintendo Entertainment System. Before these systems, we were content to play the many joystick-slapping arcade games in 7-11s around the neighborhood. Now we have the ability to pack an entire arcade experience into our homes, often into one machine. This opens up a whole new video-equipped gaming world.

In this section, we will take a look at the top-selling video game consoles, what they are capable of doing, and what to look for when you decide it's time to buy (for your kids, or covertly, for yourself).

VIDEO GAME CONSOLES

TERMINOLOGY AND DEFINITIONS

Console. Also known as the video game console unit, or the base system. This is the main part of the video game system, into which you plug the power, joystick and games.

Hardware. This consists of the physical devices of the game system: joysticks, console and game cartridges.

Software. This is the program contained in a ROM cartridge or CD-ROM. The cartridge itself is hardware, but the actual *game* programmed into the cartridge is software.

CPU, Central Processing Unit or Processor. The CPU is the computer chip used to run the core of the system. It is similar to a computer's CPU.

Cartridges and CD-ROMs. The game software is stored on either CD-ROMs or cartridges. Cartridges use memory chips (ROM), which is fairly expensive at this time (but fast). CD-ROMs can store mass amounts of data and cost very little (but run comparatively slow).

Bits, Bytes and Binary. Bits, bytes, kilobytes and megabytes relate to the memory size or capacity of the console unit's memory or its CPU. A *BIT* (b) is the smallest unit, either a zero or a one. The higher the number of bits, the larger the number it can represent. A *BYTE* (B) is eight bits and can represent a decimal number between 0 and 255. Adding one additional bit doubles the number you are able to represent; in other words, 9 bits can represent a number between 0 and 511, etc.

Bit and *byte* are typically prefixed with a K or an M. *K* or *KILO* is one-thousand twenty-four (example: 6k equals 6144). *M* or *MEGA* (meg) is just over one million. When dealing with consoles, a *meg* is usually one million *bits*. So, a 16 meg cartridge actually has two mega*bytes* of data stored onto it (16 million bits divided by 8). A CPU is measured by the number of bits it can handle, such as a 32-bit CPU. This means the CPU can handle one 32-bit number at a time.

BASICS

Look for the number of bits of the CPU while shopping for a game console. A 64-bit system is today's standard. Pay attention to the hardware it uses, such as a CD-ROM, and the costs of adding hardware and games. Can the system be expanded? At what cost? By buying a name brand, you are ensuring that you can buy many quality games for the console. Let's look at a short list of popular manufacturers:

Nintendo. This company, strangely enough, started as a manufacturer of card games. Back in 1985, Nintendo created what many would call the king of all game consoles, the Nintendo Entertainment System (NES). *Super Mario Bros* was their first game. It was a near carbon-copy of the arcade version of the game.

> **NOTE**: When you hear of an *8-bit* system, or of the newer *64-bit* systems, realize that these numbers represent the maximum number of *bits* the CPU handles at once. The more bits it can handle, the faster it works and the better the quality of graphics and sound.

The NES held the video game spotlight for many years. It still remains a favorite even though it is technologically inferior by today's standards, because of its 8-bit CPU. Hundreds of games were created for it.

After the NES came Nintendo's hand-held 8-bit version, the GameBoy. This unit used a poor-quality LCD display that was able to reproduce limited graphics. However, people wanted it! As a result, many other companies developed games for GameBoy. After creating the Super Nintendo Entertainment System, Nintendo developed a plug-in Super GameBoy unit so GameBoy owners could play their games on the TV.

The best system on the market for years was the Super Nintendo Entertainment System (SNES). The 16-bit system handled hundreds of on-screen colors, had a great sound system, and had about 750 games available. SNES came with specialized graphic function chips inside the console which also came as add-ons through special cartridges. The SNES is still for sale at around $100, and typically comes with a free game cartridge. If you're looking for a respectable, cheap console and games for whiling away the hours, I recommend one.

The introduction of the *Nintendo 64* (N64) blasted us with tens of thousands of on-screen colors, an awesome stereo sound system and textured polygon graphics. The Nintendo 64 uses a 32-bit CPU developed by Silicon Graphics to boost the three-dimensional graphics capabilities. To make the graphics lifelike, the N64 uses polygon graphics techniques to create a 3D world inside the games, complete with shading and lighting. This makes the game objects resemble real-world objects.

After saving the fair princess from the clutches of the evil Turtle Lord (in all their 3D glory), you may find a new favorite console in the N64. This system will boggle your mind with its effects. Not many games for the N64 have been released as of this writing, but soon we'll see more visually exciting adventures. The great news is that the N64 costs under $150! At this time, no free game is included. Games are currently upwards of $70 each.

Sega. Sega released its own 8-bit system around the time the NES came out: the Sega Master System (SMS). It was a great console with many games available. You could play games either with cartridges or with a tiny game card plugged into the front of the console.

One of the best peripherals for the SMS was a set of 3D plug-in glasses. They had a see-through LCD in each lens which alternately blanked one eye at a certain rate. The TV screen was synchronized with the glasses in order to give a very realistic 3D effect.

The hand-held system called Sega Game Gear (GG) was released years later. You could plug it in and play almost all your SMS games with a special adapter. This meant you could play your expensive SMS games on the run. What a wonderful, mobile feature! There were around 88 games made exclusively for Game Gear.

Amazingly, Sega later created another portable game system, the Sega Nomad. It has the abilities of the Sega Genesis, with all of its power contained within a small unit you can hold in your hand.

In 1989, Sega released their 16-bit game system, the Sega Genesis. See Photo 7-1. It could handle less than a hundred colors on the screen at a

Photo 7-1. The Sega Genesis. Reproduced with the permission of Sega of America.

time without special programming (which game companies rarely used). It also featured digital sound and a stereo music synthesizer. With a library of over 500 games, this system could keep you busy for years.

Not long after that, Sega released two new devices for the Genesis. The first, in 1992, was the Sega CD-ROM unit, which added the ability of near full-motion playback, real CD audio and massive storage space for sounds and more, larger games.

Next came the Genesis 32X. It had two 32-bit RISC processors and a newly designed video digital processor chip for a fast processing speed, more than 32,000 colors, texture mapping, improved computer polygon graphics technology and 3D graphics. With the release of the CD-ROM and the 32X, there were about 57 more games released for the Genesis.

Sega's newest system, the Sega Saturn, hit the shelves in early 1995. See Photo 7-2. The Saturn uses two SH2 32-bit RISC processors, a double-speed CD-ROM drive, and two digital video processors allowing for 16 million colors and 3D graphics. The sound system utilizes CD-ROM audio, 32-voice/3D digital sound, and a Dolby surround sound system. There are currently around 159 games out there for the Saturn.

Photo 7-2. The Sega Saturn. Reproduced with the permission of Sega of America.

For the smallest (and youngest) game users, Sega created the Sega Pico. Kids can use a "magic pen" to trace along a large stylus pad, with the results appearing on screen. A small range of about 18 titles are available, ranging from play-along stories, games, drawing and learning tools.

Recently, in a giant leap of gaming technology, the Sega Channel was born via cable TV. It brings a world of excitement into your home by allowing you to download Genesis games over your television via a special hookup to your local cable company. The games stay in memory until you're done playing or until you turn off the unit. This service usually runs $11.00 to $20.00 a month to use.

Sony Playstation. What seems to be a serious console contender (outside of Nintendo and Sega) is the 32-bit Sony Playstation. See Photo 7-3. It has enhanced abilities for three-dimensional polygon graphics (which is becoming normal for current systems), a great sound system, a CD-ROM drive that can also play audio CDs, and about 150 available games with many more on the way.

Sony is also offering wonderful incentives to game developers, such as the ability to link the system to a PC. This will let programmers de-

Photo 7-3. The Sony Playstation. Reproduced with the permission of Sony Computer Entertainment.

velop games more easily. Maybe this means that we will eventually see some of these new creations stamped onto CDs for purchase.

CONSOLE ADVANCES

3D. The next generation of video game consoles is now moving into a three-dimensional world. The antiquated 2D game environments are being replaced with consoles that can do the needed calculations to simulate a 3D world on your TV screen. Wouldn't you like to move your game characters in textured, rotating landscapes instead of simply up, down, left and right? When Sega released the Saturn, and the Sony Playstation came along, companies were able to take the 32-bit technology to a new level by creating 3D environments within a video game. Now, with Nintendo's breakthrough 64-bit engine, this new 3D world will no doubt expand even further. Game companies are still discovering the little tricks that these console machines can do.

CONSIDERATIONS

A lack of available games for a console is often a problem. Many systems have flopped because people refuse to purchase consoles that have only a handful of games available. Newer systems are not offering loads of software that will run on them. It seems that the console wars may face a tough time when companies start to put their efforts elsewhere on the market. This game-title lull is typical when new systems come out; it has happened before, and will continue to happen.

Another problem with consoles is one familiar to computer owners: equipment outdating. You can purchase a system, and almost by the time you get it home, it is outdated by a flashier, more awe-inspiring one. In the race to produce better systems, companies are constantly outdating the older ones. This stops further games from being released for certain systems. After all, what company wants to develop games for an obsolete console?

What's the solution to these two problems? Only to buy proven systems with plenty of games available. Look for a console that either has a lot of available games or has a continuous flow of newly-released games. Save the flashy new system with only three available games until later.

GAMES

Many types of games are available to suit every taste. See Photo 7-4. Let's take a look:

Action Games. These have the player running, shooting and solving simple puzzles, all while saving the world from utter destruction. Fast action gives way to fast reflexes, which are needed to win. Most console games fall into this category. Here are a couple of examples:
    Super Mario Bros. (Nintendo for NES.)
    Sonic the Hedgehog. (Sega for the Genesis.)
    Clockwork Knight I & II. (Sega for the Saturn.)
    Bomberman. (Hudson for most systems.)
    Panzer Dragon Saga. (RP for Sega Saturn.)

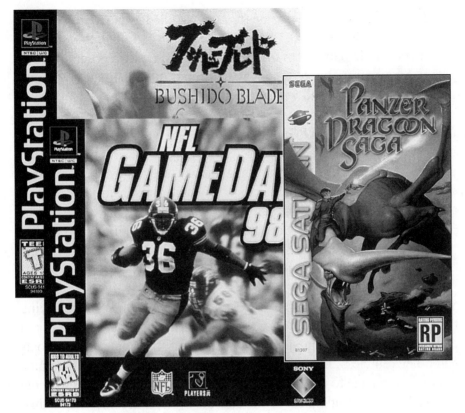

Photo 7-4. There are a variety of video game cartridges and CDs available, covering a wide range of tastes and themes. Games are produced and packaged for specific consoles and cannot be interchanged between different console makes. Reproduced with the permission of Sega of America and Sony Computer Entertainment.

Adventure Games. These take the player along a quest, solving a series of puzzles and working the mind to the extreme. Although not many are available for game consoles, there are popular games like *Kings Quest*. A few more games fit into this category:

Alone in the Dark. (T*HQ for the Saturn.)
Tomb Raider. (Eidos Interactive for the Saturn.)
Bushido Blade. (Square Soft for the Playstation.)

Fighting Games. These games pit players against one another in a battle to the end. Jumping, punching, kicking and blocking make for some great times. There's no shortage of fighting games for all systems, with the most well-known being:

Street Fighter Series. (Capcom for most systems.)
Virtua Fighter. (Sega for the Saturn.)
Mortal Kombat. (Midway for most systems.)
Tekken Series. (Namco for most systems.)

Chapter 7: Video Game & Internet Consoles

Puzzle Games. These video games require knowledge and a well-worked mind to accomplish a goal. An example is Tetris, in which you stack objects in certain patterns to lower the stack and rack up points. Puzzle games are usually simple games with simple rules that suck you into them for hours, days and weeks at a time.

Role Playing Games (RPGs). These games go way back to the days of Dungeons and Dragons, when groups of people sat around assuming the role of an alter ego. In their imaginations, they journeyed in search of fame and fortune, with monsters to slay and damsels to save. Here's a sample of some great RPG games:
    Final Fantasy Series. (Squaresoft for the NES, SNES.)
    Phantasy Star Series. (Sega for the SMS, Genesis.)
    Ultima Series. (Origin for most systems.)

Simulator Games. These take the player high above the ground, flying the fastest jets, the oldest replicas, and the most powerful machines out there. These games simulate the physics of jets, helicopters, cars, boats and other vehicles. There are a few simulator games for console units, but most are made to run on computers. This will soon change with the new 3D polygon graphic chips that console systems are now using. Just to name a couple:
    Ace Combat 2. (Namco for the Playstation.)
    F-22 Interceptor. (Electronic Arts for Genesis.)
    Starfox I, II & 64. (Nintendo for SNES, N64.)

Sports Games. These games bring out the armchair athlete in you. There's much punting, dodging, jumping and grunting while your onscreen ego shoves through a line of fullbacks to make that 6-pointer. Sports games make up a good section of video games out there. There are too many sports games out there to name the best, but here are a few:
    John Madden Football Series. (Electronic Arts for most systems.)
    E.A. Sports Series. (Electronic Arts for most systems.)
    NBA Jam. (Acclaim for SNES, Genesis.)
    NFL GameDay. (For the Sony Playstation.)

Strategy Games. Most strategy games test your "general" abilities in warfare: deploying tanks and soldiers, and taking over villages with strategic locations and such. Other strategy games like *Tetris* are in this dual-category because it requires a lot of strategy to foresee and plan your moves to win. Some examples:

    Tetris. (On most systems.)
    Lemmings. (Psygnosis for most systems.)
    Populus. (Electronic Arts for most systems.)

WHERE TO GET GAME CARTRIDGES AND CD-ROMs

Buy. Look for the best deal. Most new or popular games are VERY expensive. However, sometimes a gem can be found in the reduced-price bin at K-mart.

Trade/Borrow. Most gamers trade cartridges with each other either permanently or for a few weeks. Maybe a friend (or even your kid's friends) has a system like yours.

Rent. Video stores are responding to consumer demand by renting out game cartridges and CD-ROMs. Most rent Nintendo, Sega and Sony Playstation games for a few dollars a day.

CONSOLE EMULATION AND COPIERS

Computer users have emulated (simulated with a different device) their favorite video game systems on their computers for years. The idea is to load console games onto a personal computer to play them. Sound and pictures are sometimes better on a PC. An added bonus to doing this is the ability to modify or create your own games!

There are several great emulators out there for the Nintendo GameBoy, NES, SNES, Sega Master System and Sega Genesis: enough to keep you busy for hours. By doing a search on the Internet, you will be able to find what you are looking for and contact the right companies from which to purchase an emulator.

NOTE: It is against the law to download any unpurchased games for emulators. Using illegally duplicated software is a violation of copyright laws. You must own the software, and by law, you are allowed a backup copy as long as you still have the original software.

How do you get a cartridge game to your computer? You can copy the contents onto a computer disk by plugging your cartridge into a reading device. These "copiers" are either stand-alone units or can be hooked to your computer. The most popular stand-alone units act as disk drives, of sorts. A stand-alone unit stores a game on a floppy disk so you can load and play it on your game console: they plug into the cartridge slot of your system. Stand-alone units boast such features as game saving and codes to alter the game. They also have the ability to allow you to alter and create your own games which can then be loaded up into your game console!

HOOKING UP YOUR SYSTEM

Hooking up your new system shouldn't be hard (yeah, right). There are two basic types of connections. See Figure 7-1. One is connecting the console through a typical RF box (a small box with a short coaxial cable, a female coaxial connector and a long thin wire). This is most commonly used hookup method. The second is through A/V cables. This splits the video signal and the audio using two or three lines (one for video, and the other one or two for audio).

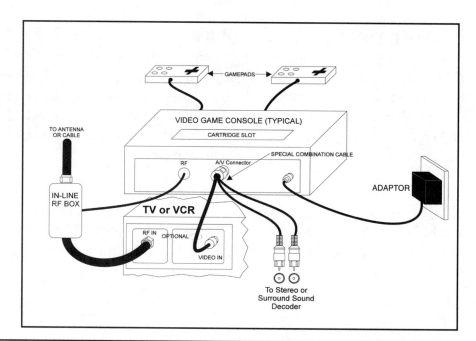

Figure 7-1. Hooking up a video game console.

You must select a channel that is unused in your area when hooking a video game console to a TV or VCR via an RF box. The system will convert the signals to that station to display the video game on the TV screen. If the console is hooked up through your TV, put on channel 3 or 4 and match the channel selection to your system. If hooked up through your VCR, make sure the VCR is set on the correct output channel for the game console, and that the TV is set on the channel used by the VCR.

The RF box connects in-line with your TV or VCR by way of the co-axial cable. The antenna or cable coaxial attaches to the female side of the RF box. The coaxial on the box itself connects to your TV. Lastly, simply plug the long thin wire into your game system.

A/V lines split the video and audio into separate cables so you can plug them into your TV or VCR's composite video jack and the stereo's audio jacks. If there is a third cable, the system is stereo and should be connected accordingly. Most TVs and VCRs have auxiliary inputs and settings to accept video and audio signals, instead of switching to another channel. By hooking the video console through your VCR, you can record parts of the game being played: the ultimate proof that you did indeed win the game. Set your VCR to the appropriate channel for the game system and hit *Record*. Presto!

Next, plug your power adapter into the back of the system and into the wall socket. Make sure to hook the joysticks into the correct ports. Put in a game and go for it!

TROUBLESHOOTING

Typical problems which plague game consoles are the adapters, video connections and joysticks. Take care you do not bend any wires, especially the power supply's. Right where the wire enters the adapter case is a troublesome area: the wire often becomes broken or frayed inside of its plastic tubing, cutting the console's power. All of my adapters have messed-up wires that need to be twisted and turned just right to work correctly. Faulty adapters usually have to be replaced as they are sealed and usually unrepairable.

Another problem occurs when cartridge and console connectors collect dirt and dust, causing a bad connection. Most systems have a cleaner available, or you can use a good solvent and a Q-tip to clean each connection. Rubbing alcohol works fine. Just make sure the solvent is dry and any loose particles are blown off before plugging the game back in.

The main problem with joysticks is *sticking* joysticks or buttons. If you feel proficient enough, open them up and clean the casing and connections with rubbing alcohol, paying special attention to the contacts. This usually clears up any problems. If not, check all of the connections to the console. When all else fails, replace the joystick; they are cheap nowadays, anyway.

Check your user's guide for more information on troubleshooting consoles and their hardware. You can also email or call the company's technical support staff for more guidance. Remember, opening any console or joystick components may invalidate your warranty. Check the warranty papers that came with the console. (You DID keep them, right?)

INTERNET CONSOLES

The Internet's growth has exploded in the past few years. What was once a bunch of students and academics sending email and educational-type electronic information back and forth has turned into an entirely new communications medium.

With the birth of the World Wide Web came a tremendous upsurge of people getting on-line. Email and Web pages have become as common as phones and faxes. The problem used to be that you needed an expensive computer to access the Internet. Not anymore. Internet *appliances* are just beginning to be released. These are TV set-top boxes you hook up to your TV to surf the Internet or send email across the globe, all via remote control or infrared keyboard. Now you can explore the Internet from your couch.

Why the word "appliance"? It's basically due to the fact that companies want to make Internet consoles as common and affordable as any other appliance in your home.

Types of Internet Consoles. As the product is so new, companies are trying many different style formulas. Here are the three ideas manufacturers have in the works for Internet consoles:
1. A stripped-down personal computer. This is to try to keep the cost below $500.
2. A stand-alone set-top box connected to your television. A typical computer monitor costs $250 to $1000, so the companies are trying to get rid of that cost by integrating the Internet console into your home entertainment system.
3. A special Internet peripheral plugged into a video game console so you can use it to access the Web and email. (Cost is around $200 for the peripheral only.)

TECHNOLOGY

Most of these Internet devices use a high-end microprocessor, 2 to 64 megabytes of RAM and a modem. What else? Not much! The idea behind these simple machines is to strip the parts count down, thus driving down the cost. The downside is that you can't save files for later use, something computer users take for granted. However, some units allow you to connect to a printer for hard copies. Does this sound so bad compared to a $2500 computer? For the simple functions it provides, an Internet console may be all you need at first.

Some Internet consoles make use of a programming language called Java. This computer language, designed by Sun Microsystems, Inc., works on nearly any computer platform such as UNIX, Windows, OS/2, MAC OS, etc. There are no compatibility problems with Java programs. A Java word processor designed on a Macintosh would run equally well on a Windows '95 system.

Because Java can run on any system, it is now possible to control any electronic appliance with it. That, of course, includes Internet consoles.

THE DIFFERENCE BETWEEN COMPUTER MONITORS AND TVs

Here are some common questions in the field of TVs and computer monitors: "What is the difference between a TV and computer monitor?" "Why do monitors cost so much more than a TV?" "Why can't I just use a TV for my computer's monitor?"

The main difference between a TV and a computer monitor is the resolution (scan lines). A TV reproduces a low-quality, unsharp picture. An NTSC signal is only 525 vertical scan lines and approximately 300 horizontal scan lines.

Computers need high-resolution monitors to reproduce fine details for graphics. A computer's resolution is measured in horizontal-by-vertical pixels (dots that make up the image). Today's monitors can display a 640 pixels by 480 pixels (640x480), up to 1600x1200±. This is twelve times more picture information and a much sharper an image than a TV can produce. It is almost like looking at a photograph.

If you have ever looked at text on a TV screen, the resolution difference is obvious. If the text is anything under an inch tall per letter, then it is just too fuzzy to read. Annoying!

Another difference between a monitor and a TV is that a TV uses an interlaced scheme and monitors do not. Interlacing cuts a picture into 250 odd horizontal lines and 250 even horizontal lines. It first displays the odd lines of the picture every 60th of a second, then displays the even lines the next 60th of a second. In this way, a TV creates the illusion of one complete picture.

A computer monitor does not break the image into even and odd fields, but instead displays a complete image each 60th of a second (non-interlaced monitor). This helps reproduce motion, such as an object moving across the screen, with no perceivable flicker. TVs cannot do this if the object moves too fast across the screen.

The result of higher resolution and non-interlacing in computer monitors is higher-priced tubes and circuitry. You pay for the monitor's ability to reproduce fine images. However, some TV manufacturers are starting to produce higher-resolution TVs that mimic a low-resolution computer monitor quite well. These can be used as an alternate monitor or hooked to an Internet appliance.

For more information on Java, see the Sun Microsystem, Inc., Website at http://www.sun.com/java.

INTERNET APPLIANCES CURRENTLY AVAILABLE

Only a few of these devices are currently on the market in North America. Here is a description of each:

WebTV Networks. WebTV is a technology licensed to electronics manufacturers. Philips/Magnavox and Sony currently produce a set-top box that attaches to your television. See Photo 7-5. The WebTV Network or one of their partners provides the service that you use to access the Internet. Contact WebTV Networks for more information at http://www.webtv.net or 1-800-GOWEBTV.

Philips/Magnavox and Sony WebTV TV-Top Internet Appliance. WebTV units enable you to plug the set-top box into your TV. All manufacturer models are virtually identical in specs. They have a remote control or plug-in keyboard to allow you to surf with the pros. The current units use a 33.6 kbps (kilobits per second) modem, multiple video outputs (for RF or S-Video), and audio outputs with CD-quality sound abilities along with a MIDI sound processor. A 64-bit, 112 MHz IDT R4640 Orion RISC CPU, with two megabytes of RAM and ROM, is the brains of the WebTV.

You can upgrade the whole unit by plugging in new cards or printers. Software-wise, it uses a Netscape 3.0 and Microsoft Internet Explorer 3.0-compatible browser (using HTML 3.0 and MIME-encoded features).

Hooking up a WebTV unit is as easy as hooking the RF box to your TV, plugging in your phone line, and turning it on. See Figure 7-2. The unit will automatically dial and log you onto the server (the only server to which it can currently connect, unfortunately) and either log you in or

Photo 7-5. WebTV Internet appliance. Reproduced with the permission of Sony of Canada Ltd.

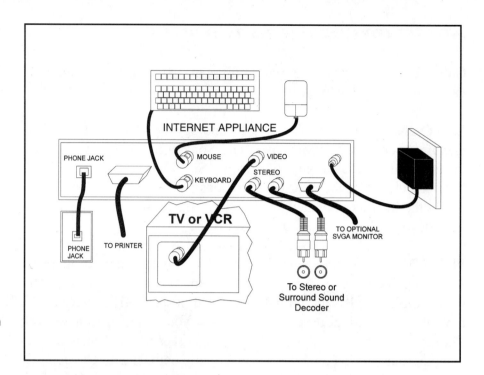

Figure 7-2. Hooking up an Internet appliance.

set up your new account. The cost of the service is about the same as other Internet providers, about $20 a month:
- Magnavox WebTV Info: http://www.magnavox.com/ or 1-888-813-7069.
- Sony WebTV Info: http://www.sel.sony.com/SEL/webtv/index.html or 1-888-772-SONY.

SEGA Saturn Netlink. Sega has created a new device to allow all Saturn owners to hook into the Internet. See Photo 7-6. The Netlink unit plugs into the Saturn game cartridge port and a phone jack. You use the Saturn controls to navigate around the screen. It uses a custom HTML 2.0 (and most 3.0 extensions), WWW browser, email and a real-time IRC (Internet relay chat) that allows for real-time communications with other people who are hooked up. The best thing is that you can use your local ISP (Internet service provider) for access! Hookup is just as easy as the WebTV.

Photo 7-6. The Sega Saturn NetLink. Reproduced with the permission of Sega of America.

NETWORK COMPUTERS

A network computer is a small computing device that has minimal hardware and runs its software mostly off of a network. It can be a stripped-down desktop PC or a miniature computer that you hold in your palm.

Network computers (NCs) are only a proposed standard for now. Apple, IBM, Oracle Corp., Sun Microsystems, Inc., and Netscape, led by Oracle NC, Inc. (NCI), have set these standards to include 640x480 resolution, mouse, keyboard and sound. The NC should run Java applications, the Internet, and boot its software off a network instead of storing it locally inside itself. You can plug a "smartcard" into the NC to access the device. This is a great security feature and can be used to adjust all of your preferred settings each time you use the NC.

When the NCs actually come out, they will have the following attributes:
- They will be architecturally neutral (can be hooked up to any system).
- They will have a much lower total cost of ownership than personal computers.

- They will have a lower entry price than a typical personal computer.
- They will be significantly easy to use and administer.
- They will enable security. (This means that no one else can access your NC.)

NCs are more than just Internet appliances in that you can also use them run Java applications such as word processors, database tools, games, etc. The wait continues for cheap computing.

See the Network Computer, Inc., Website for up-to-date information and a list of devices and available software: http://www.nc.com/.

THE FUTURE

What does the future hold for video gaming? Soon we'll see at least two more 64-bit game machines as well as add-ons for existing consoles. Most notably, there is the *64DD* (the Nintendo 64 Disk Drive). This device is a high-speed mass storage device that can handle upwards of 64 megs with an addition 34 megs writable. Another rumored feature is modem capabilities, allowing for Internet connection and multiplayer gaming.

The 64DD lets you update games with new levels, features and options. You can save your game information and take it to a friend's house to use. This will open up a new realm of console gaming that only the PC world currently enjoys.

In the distance future, we will likely see partnerships forming between gaming companies to bring us the best in gaming. One example is a possible alliance between Intel and Microsoft, who are looking to create a new breed of arcade machine that will likely trickle down to the home game market.

With the advances in integrated microprocessors and digital signal processors, we will likely soon see all-in-one entertainment packages. These would pack a TV, VCR, game console, Internet console and computer into one device.

Imagine watching a movie on your TV, a commercial comes on, and you feel like playing a quick round of the newest fighting game. So, you push a button to pop open a small window to the Net, grab a copy from the game company, and load it into the game console. Contained within its own little window, you can see when the commercial ends. Push another button, and the game closes, returning you to the movie. You could also see an advertisement on the tube that says "Visit our Website for more information!", push a button, access your browser, and automatically go to that Web address.

A complete entertainment package would to bring the entire video world into your home. It's great to be able to chat with a friend halfway across the world or join in a computer game with a hundred other people. Of course, computer users have had this luxury for quite some time, but soon everyone with a home entertainment system will be able to join in with very little fuss.

CHAPTER 8

PURCHASING HELP & RECOMMENDATIONS

"Theater is life. Film is art. Television is furniture." Unknown

How many slick sales spiels are you to endure over a lifetime? If there is a hell, it is sure to involve a timeless existence stomaching the bad tie supplies of salespeople and suffering through their technical raps.

Granted, everyone needs a job. Actually, some salespeople are highly knowledgeable in their field. However, from painful personal experience, the ones who actually know something about what they're selling are few and far between. It is up to you, the purchaser, to know the products inside and out before ever walking into an electronics store and talking to a salesperson.

This chapter is loaded with tips, hints and definitions, armor-plating to stop the barrage of bizarre business practices and win the battle of purchaser satisfaction.

CONSUMING vs. PURCHASING

Who conjured up the word *consumer*? It has become socially acceptable to trash what we use, so in that sense the word "consumer" fits: i.e., to consume, use up, spend or waste (time, energy, money, etc.). In the long run, this kind of behavior usually ends up being nothing more than a waste of money, and who wants to throw away money? The solution is to become a *purchaser* as opposed to a *consumer*.

By educating yourself and not purchasing *garbage* that ultimately breaks down after the warranty expires, you are saving your sanity, the environment, and money! Years ago items were built to be everlasting. People who purchased typically spent a few extra dollars to *NOT* become a consumer. Do yourself a favor; if the difference is a few extra dollars, opt for being a purchaser. Maybe manufacturers will eventually get the point when we refuse to invest in trash anymore.

KNOW THE JARGON

Knowing the nomenclature of purchasing, and the words of the product itself, will improve your buying decisions. Learn about the features of your intended purchase and their meaning. Chapter 9 lists common electronics features for TVs, VCRs, etc.

RESEARCH BEFORE YOU BUY

Your biggest weapon against purchasing fraud is *knowledge*. Take your time to research the product inside and out, especially if it is a big-ticket item. Try to find trade magazines with current, *independent* reviews on the product you're researching. *Consumer Reports*, *Popular Electronics*, *Popular Mechanics*, *Popular Science*, *Electronics Now*, or any of the video equipment trade magazines will probably have the information you want.

Read Ads. Ads are a valuable commodity to help you learn about electronic products. Glean as much technical info from ads as possible. While you are at it, try to get a general idea of what that item's cost is. Most stores list a manufacturer's suggested retail price: you can pretty well ignore it because in today's competitive market, the real price is usually 10-50% lower than advertised.

Compare Features. Take your time to find out what each feature *really* means and does. Do you really need to pay $100 to $500 more dollars for a picture-in-picture set?

Ask Friends. Ask your friends about the brands they've bought and the luck they've had. A friend would be the one to ask whether or not it's worth spending an extra $100 for a new feature.

Find a Knowledgeable Salesperson. They ARE out there. Look for a salesperson who does not work on commission, and pick his/her brain. Test them! Then see if the salesperson can help you understand the features you don't understand.

Get More Product Information. You can do this by visiting the manufacturer's Internet Website or by calling their 800/888 number. See the appendices for lists of Websites, phone numbers and addresses. Some stores will also supply color brochures.

PURCHASING FRAUD

The legal definition of *fraud* may differ from the moral one; but in my book, any dishonest, misleading business practice or advertising is FRAUD! Caveat emptor: let the buyer beware.

The best way to fight fraud is by researching a product to death. Find an independent review, demand information regarding a product's repair record, and insist on warranties. Most importantly, GET EVERYTHING IN WRITING so you have a case if something goes wrong.

Report disappointments, recall-quality items or downright fraud to your local Better Business Bureau or Chamber of Commerce. Here are some common things to watch out for in the home entertainment area:
- A VCR advertised as having four heads, but not four video heads.
- A salesperson (in hopes of selling you a service contract) says there is no warranty on this item. Not likely! Most consumer electronics items have a 1- to 5-year warranty.
- A TV is billed as being a PIP model, yet may not have the two built-in tuners to required to watch two channels at once. This renders the PIP function useless.
- "That model is not on display at the moment. However, this unit is *similar*." It's likely an end-of-line product. Outdated!
- Sometimes a TV's diagonal measurement is falsely inflated an inch or so by making the screens corners more square. For instances, a $600 twenty-*seven* inch TV may actually be the same thing as a $500 twenty-*six* inch model. See Figure 8-1.

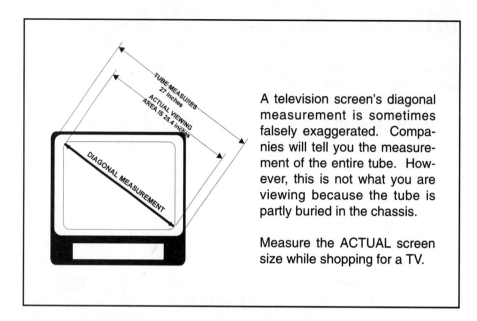

Figure 8-1. Diagonal duping.

WHERE SHOULD YOU BUY VIDEO EQUIPMENT?

There are many types of electronics stores. Which ones should you frequent? You could find what you're looking for in almost any electronics boutique, specialty store or major department store. Examples include Best Buy, Circuit City, Ultimate Electronics, Sears and Incredible Universe.

Look for stores that have these advantages:
- They stock and display the item you want.
- They service what they sell.
- Your friends have recommended them.
- The salespeople don't work on commission. This may be difficult to find, however.
- They employ salespeople who actually answer your questions and are *courteous* and *prompt* in helping you.
- They are *consistently* busy. Watch the store for a few months and see if people actually return. It's usually a sign of good prices and good service.

WHERE YOU SHOULD *NOT* BUY VIDEO EQUIPMENT

Avoid any of these types of stores:
- A fly-by-night operation tucked away in the corner of a strip mall.
- Stores that jack up prices then put items "on sale." This is a sure sign that you are about to walk into a crowd of serpentine salespersons. Watch the bite!
- Mass-marketers such as K-mart, Walmart, Target, etc., offer brands such as GE and Sanyo. These are usually considered *economy* products. If you are extremely tight on money, then by all means purchase them. If not, consider looking elsewhere as these stores will rarely offer any sales or service help for the products they sell.

The greatest advice I can give you is to shop around and look for the best deals without sacrificing the "shoulds" and "should nots" previously listed.

THE BEST TIME TO MAKE A PURCHASE

If you want to purchase high-ticket items, do it when new models come out, typically in the spring. For small to mid-ticket items, the holiday shopping season is your best bet (or shortly afterward).

PURCHASING STRATEGIES

After reading this book, you should generally know what you want to buy, be it realistic or unrealistic. You have X amount of dollars for what you want. Now it's time to balance your video budget with your video needs and wants. See Figure 8-2.

Once all of the electronics jargon is under your belt, and you are pretty sure what each feature truly does, start perusing ads in your local newspapers. Cross out the stores that meet the *shouldn't* criteria we mentioned. Go to the few lucky stores left on the list and see how the store *feels*; dishonesty has a nasty odor, and I am sure you will be able to smell it.

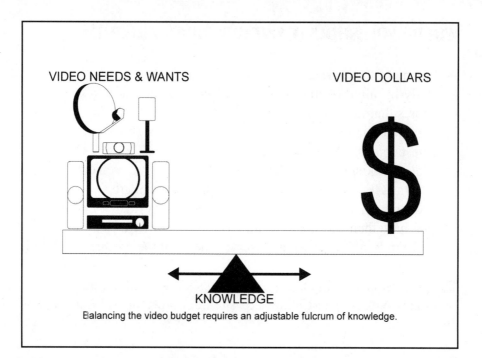

Figure 8-2.
A balanced budget.

Make a note of how fast the salespeople try to help you. If it takes more than a few minutes, don't stand around: leave. When someone does approach you, start asking questions. If you don't get your questions answered, cast your buyer's vote and walk out. No use in wasting your time, or more importantly, your money.

Before walking into a store you should know which brand you want and what to expect to pay for it. Make sure you walk out with that goal intact: take the attitude that you are there to BUY something, not to be SOLD something. The right salesperson for you will sense this and be more than willing to help. After all, it's easy money to him/her, right?

OTHER PURCHASING TIPS

- Always walk into a store knowing the brand and model YOU want to purchase. If the store doesn't have the item, try somewhere else. Stores are sure to try to sell you an overstocked or nearly obsolete item otherwise.
- Look for a price guarantee. Most electronics boutiques will pay you the difference if an item becomes cheaper within a set period of

time. Example: You purchase a VCR for $249 in August from a store with a price guarantee. The same store runs a $199 special in September for the same VCR. The price guarantee allows you to go to the store with the receipt and get a $50 refund.
- Find an establishment that covers the product's warranty on the premises.
- Beware of service contracts. The best alternative is to have a "slush fund" put aside for all of your appliance repairs.
- Become a purchaser, not a *consumer*.

PURCHASING RECOMMENDATIONS

Now that we have covered purchasing strategies and pitfalls, let's turn to a more upbeat subject: tips and recommendations on selecting a brand and model.

IMPROVING YOUR SYSTEM

Evaluate Your Current System. In Chapter 2 we discussed sights, sounds and signals. By evaluating each of these areas, you will be able to see where your system needs to improve.

Sight: Are you starting from scratch? Look at your TV as the first step. Is it large enough for your visual tastes? Will it support high-resolution equipment such as DVD?

Sound: Do you really want surround sound, or will a simple stereo upgrade suffice? Can you add a larger amplifier or a special effects A/V receiver to satisfy your discerning ears?

Signal: Are you content with local channels, or do you want million-channel cable service or a DBS system? Can the signal be improved with an amplifier or another piece of hardware?

What Improvements Can You Make Within Your Budget? By adding one component at a time, you can spread the high cost of updating over a period of time. The first heavy investment you should make is in screen size and resolution. After this, it is a matter of adding a VCR here or a DBS there, and sneaking in an A/V receiver with rock-the-house speakers. Eliminate a few expensive features (which are usually useless anyway), and you'll be able to buy that new camcorder this week instead of two years from now. The whole idea is to build the castle one brick at a time.

What's the Plan? Lay out your ideal system, and be somewhat reasonable about what you want. Try drawing a diagram with the various components (TV, VCR, etc.) and list the features you want for each. Do some research and see which brand matches this criteria. Look for a brand name with a good reputation and see what they have to offer: visit Websites, ask for brochures, and make inquiring phone calls to the manufacturer or district representative.

Does your budget allow for the model that has all the bells and whistles? Probably not. Most companies offer a reduced feature model for a slightly lower price. Is it worth waiting for the extra $100 in the budget, or can you do without a particular feature? An example would be a camcorder: you want a color LCD screen, but the cost difference between that and the black-and-white LCD screen is $200. Is it really worth it?

Resolution, Digital. Put most of your video investment into a TV's video resolution. Get at least a 500-line set so you can enjoy the benefits of DVD and other high-resolution media.

Any new digital video equipment is a good investment as most of the current and upcoming video components will use digital techniques. It is worth a small additional investment on a high-ticket item rather than risk having to replace the whole item in the near future.

Cheapest Model or Quality Model? Remember the saying, "You get what you pay for"? Some people walk into an electronics store and simply want the cheapest possible model, forsaking any quality or helpful features. These models are usually stripped down components, or are going out of production, or are overstocked. What usually happens is that a newer model is about to be released, and the stores need to get rid of the old junk first. Try to avoid these "deals" if you can. Instead, be a comparison shopper and find the *sweet spot* for that piece of equipment: what are the best features to cost ratio. For instance, if you are looking for a VCR, a cheap model will run $150; but for $50 to $100 more, you can get one with hi-fi, four video heads, front jacks, VCR Plus+, and a heck of a lot more. Find each brand and model's sweet spot and remember, "Twice the price equals twice the quality." In some cases, more!

Avoid Duplicate Functions. Try to avoid buying equipment with features you already have on another piece. For example, if your TV came with a universal remote, don't waste the extra cash for one when buying a VCR.

Complexity of Equipment. Don't buy complicated devices unless you know or can easily learn how to use them. By looking at the owner's manual, you can learn something about the unit. If the manual is well-written, the unit will likely be simple to operate. By performing a few operations from the manual, you will be able to tell how complicated the functions really are. Some manufacturer on-screen programming sequences are infamous for being complicated. While shopping, test a few functions to determine their complexity and usefulness.

See the Equipment for Yourself. Most large electronics boutiques have sound rooms: nothing beats seeing and hearing the equipment for yourself. Ask for a demonstration and for literature on each device that seems interesting. Beware, though: some places only display high-end equipment in special sound rooms. If the model you want is not set up, kindly thank the salesperson and find a place that does have it set up.

TELEVISION SET

Size. Get the largest possible screen your budget (or your living space) will allow. See Photo 8-1. Upgrading from a 27" to a 32" tube may not seem like much, but you add 40% more viewing area. A 37" would more than double your screen's viewing area.

Resolution. Pay attention to the horizontal line measurement of the tube. Get at least a 500-line model so you can take advantage of a

Photo 8-1. A 61" rear-projection TV. Many videophiles dream of owning a screen like this but lack the space. Buy the largest screen you can afford, but also make sure your room is big enough to allow you to watch the TV comfortably. Reproduced with the permission of Sony of Canada Ltd.

digital video camera and digital video disk player. If you are planning a system intended to last seven or eight years into the future, then get a 1000-line or higher model for HDTV purposes.

Jacks. Make sure the set has A/V and S-input jacks. Get front jacks if possible.

Sound. You should get a minimum of MTS stereo and preferably surround sound of some sort.

Misc. Forget PIP if at all possible. I guarantee that you will use it once or twice and wonder why you spent a fortune for this feature. If you are purchasing anything over a 25" model, make sure it has comb filter circuitry. Also, make sure the TV comes with a universal remote and simple on-screen programming.

A/V HOME THEATER RECEIVER

Stereo or Surround Sound? Do you want surround sound or will a cheap MTS stereo unit suffice? Does your TV already have MTS/SAP built in? If so, avoid the equipment duplication. If you opted for surround sound, buy at least a Pro Logic receiver and two or three quality speakers. You can switch to the Dolby 3 mode for the time being until you are able to add surround speakers and a subwoofer.

Wattage. An amplifier is measured in watts per channel, and is usually built into the A/V receiver itself. Most advertisements and spec sheets list the overall wattage or watts per channel. For example, an ad will show 320 watts total or 80 watts x 4 channels of equal power (which means the power is divided between four channels, equally). The three front speakers use about 75% of the power and the two surround speakers draw about 25%. Some models will have a separate figure for subwoofers, which pull loads of power to give thunderous bass effects. If you want a low- to mid-range amplifier, look for 150-300 watts total or 50-80 watts per channel. Mid- to upper-range is 400+ watts total or 120 watts per channel with a 100-watt subwoofer. The cost of the amplifier/receiver will be about $1.00 to $1.50 per watt (e.g., $300 for a 250-watt unit, or $600 for a 600-watt one).

Speakers. Speakers range far in price and quality. Your best bet is to examine independent studies and listen to a few soundtracks in an electronics store "home theater" room. Most stores have switchable speaker sources to let you hear different models and brands. Make sure you listen to familiar tracks to compare quality.

VCR

Hi-Fi or Mono. Try to spend the few extra dollars for hi-fi. Most VCRs are hi-fi these days, so the cost is continually coming down. Mono VCRs will soon be throwaways.

Jacks. Look for a model with an extra set of A/V jacks on the front panel so you can easily plug in your camcorder for dubbing.

Video Heads and Features. Expect to pay $150 to $300 for a plain, tasteless two-head VHS VCR. If you are on a budget and don't particularly see a use for pristine still images or knock-your-socks-off sound, then go for a mass-market, cheap VCR. However, in the $200 to $400 range, expect many more flick-enhancing features such as four to seven video heads, hi-fi stereo sound, 181-channel tuner, digital audio tracking, on-screen programming, VCR Plus+ or StarSight and (get this one) an automatic head cleaner. My favorite feature is the commercial skipper!

CAMCORDER

Format. Decide on the format you wish to purchase. The basic choice these days is VHS-C or 8 mm. VHS-C is convenient as you can play the tapes on a VCR with an adapter. However, 8 mm is extremely popular these days because of its longer recording time and better quality. The second question is, "Do I need a high-resolution camcorder such as a S-VHS-C or Hi8?" A high-resolution camcorder is nearly twice the cost of a regular 8-mm or VHS-C camcorder. Maybe you want to make the digital investment and go with DV format. Expect to spend three to four times the cost of an 8 mm or VHS-C.

Viewfinder/LCD. Try to buy an LCD viewfinder camcorder because the advantages of viewing in the field are worth it. The current viewing trend for camcorders involves small, flip-out color LCD screens. Decide if the extra cost is worth it. If you want to save a few dollars, go for a black-and-white screen.

Features. This is where your shopping skills will come most in handy. Some camcorders load up on features but still run the same price as bare-bones camcorders. Compare, compare, compare!

DBS

DSS or Other DBS Services. Take a serious look at the services and channels you want to order. Go to the DirecTV and USSB Website for DSS programming, or go to the Primestar or Dish Network Website to compare channels and features.

Equipment. Decide if you want to make an initial equipment investment or if you would just like to rent the unit.

Installation. Try to get a reduced installation fee if you purchase your system.

RATING A BRAND

When push comes to shove, brands usually don't matter: it's the model differences that count. However, when going shopping for new equipment, you may want to rate brands and models yourself. Try to rate with these criteria:

Television. Picture, resolution, sound signal reception, ease of on-screen controls.

A/V Receiver. Sound quality, power, ease of controls.

VCR. Picture, ease of programming and use, features per dollar.

Camcorder. Picture color and clarity, sound, features per dollar, editing capabilities.

WRAP UP

Throughout this book are terms, explanations, lists of features and advice. Hopefully by the time you walk into a store, you will know what you want and how much you are likely to pay. You probably will have more technical video knowledge than the salespeople! Good luck, and happy shopping!

CHAPTER 9
FEATURES

"Don't you wish there were a knob on the television to turn up the intelligence? There's one marked 'Brightness,' but it doesn't work." Gallagher

Once you have figured out what you want in the TV set, VCR or other piece of video equipment that you plan to have delivered to your doorstep this week, it is time to start deciding on features. Listed here are a few more features in addition to the ones listed in previous chapters.

Most manufacturers list model features on their Websites and other promotional materials. Also, look for reviews or comparisons of items to save yourself time in looking for the features you want.

TELEVISION

Black Level Expansion. Increases the picture tube's contrast by making brighter whites and stronger blacks. Also called *black stretch*.

Built-In Tuner and/or Cable Converter. Some TV sets come cable-ready, eliminating the need to rent cable boxes. The only drawback is that some cable companies require you to use their cable converters in order to receive premium channels.

Comb Filter Circuitry.* Enhances a picture by increasing its resolution and eliminating extra color in the picture's detail.

Dark Screens. The manufacturer slightly tints the TV screen. This reduces glare when the TV is in a brightly-lit room, and increases contrast.

NOTE: Features with a "*" next to them are ESSENTIAL or important features.

Flat Screens.* This feature reduces glare and stops the picture from distorting around the edges of the screen. If at all possible, get a screen with this feature: the difference is amazing.

Instant Replay. Lets you see the last few seconds of what was just viewed. (Only on top-of-the-line models.)

Jacks.* Connections from which the signals are sent or received. Look for sets with multiple input sources such as A/V, S-Input, RF and speaker jacks, if the TV has surround sound. Most newer sets now have front jacks for the convenience of plugging in video game consoles or camcorders.

Light Sensors. Automatically adjust the TV's brightness and contrast according to the lighting in the room.

Lines of Resolution.* Usually refers to the TV set's horizontal resolution. Look for a TV that is capable of at least 500 lines if you are going to expand your home theater in any way.

Picture-in-Picture (PIP) and Multiple Tuners. Lets you watch two programs on-screen at once. Beware! In order to watch two shows simultaneously, you need a TV set that comes with two built-in tuners; otherwise, the PIP can only work with a video input device like a VCR or camcorder.

Sharpness Control. Hones the picture's clarity and helps eliminate snow caused by being too distant from a broadcast station. You can also use it to adjust the picture's quality when switching from TV to VCR signals.

Video Noise Reduction (VNR). Smooths the edges of images hit with noise problems. Similar to sharpness control but much more complex.

SOUND

Closed Captioning. All TV sets produced after July of 1993 are required to have this hearing-impaired feature. Some sets even turn on this feature automatically when you hit the *Mute* button.

Dolby Surround. See Chapter 2 for a detailed explanation.

Enhanced Stereo. See Chapter 2 for a detailed explanation.

Front-Firing Speakers. The TV's internal speakers are aimed at the viewer, making a better stereo effect. It's better than *side-firing speakers* because it reproduces a center channel effect much better.

Separate Audio Program (SAP). Some broadcasters send a separate audio track for the same channel. For someone whose language is other than English, SAP allows programs to be received with an alternate language soundtrack.

Side-Firing Speakers. By placing the speakers at the sides of the TV set, manufacturers try to enhance the stereo sensation. However, if the TV is placed in a cabinet, the effect is negated.

Speakers On/Off. Lets you turn the TV's internal speakers off when you make use of an external audio system.

Volume Leveling. This "levels" a TV's sound. In other words, it automatically lowers strong, annoying sounds and raises soft sounds.

HOME THEATER A/V RECEIVERS

Front Speakers, Center Speaker, Surround Speakers, Subwoofer.* The speakers used to listen to surround-sound stereo.

Dolby 3 Mode.* Simulates surround sound with only three speakers.

Dolby Surround, Pro Logic, Digital (AC-3) Decoder. Decodes the various Dolby surround sound signals for playback.

Digital Signal Processing (DSP). A microprocessor used to process, correct, manipulate or modify a musical signal using digital means.

On-Screen Display. In conjunction with the remote, it allows you to control the receiver's functions from the comfort of your chair.

Power Output. This is the power provided to the speakers by the amplifiers. It's measured in watts per channel.

Random Preset Memory. Lets you preset AM or FM stations for easy recall.

Sound Field Patterns. With the push of a button, you can simulate the acoustic patterns of an arena, dance hall, jazz club, hall, live club, pavilion or theater.

THX Home Theater. Accurately recreates the experience of a movie theater sound system.

Tone Controls.* Control the bass and treble or graphic equalizer settings. Some models allow separate control of the center dialog channel.

REMOTE CONTROL AND ON-SCREEN PROGRAMMING

Active-Channel Scan. Allows you to flip through the channels without lifting a finger.

Alarm/Sleep Timer. Can be used as an emergency wake-up device. Nothing like a $700 alarm clock; but it has other uses. You can set the TV to come on for a few hours at night when you are on vacation to deter burglars from stealing your new TV!

Channel Auto-Program. This is the TV's PLL at work. It locks onto all of the available channels so you can easily flip through them in the future.

Channel Block. Great parental guidance feature. You can assign certain channels to be accessible only by code. You will have to learn how to do this one yourself. I guarantee that the kids won't help.

Channel Labeling. Can be used to name a channel. Great if your TV guide has labels and not numbers. For example, you can program the channel X TV display to read "HBO."

Clock.* This is the internal clock used to sequence events or displays for the user.

Channel Reminder Display. An unobtrusive, on-screen display of the current channel. Can be left on continually or set to show every 30 minutes or so.

Commercial-Skip Timer. A great commercial eliminator. Flip this feature on, and you can surf during a two-minute commercial. When the two minutes are up, you will be returned to the original channel.

Digital Control.* Controls all of the TV's functions with a microprocessor instead of with external dials and buttons. Nearly all TVs currently being produced make use of this.

Multilingual Menus. Displays the on-screen programming menus in different languages.

On-Screen Menus.* Nearly all new TV sets have this feature. It lets you see all of the TV's controls on the TV screen, such as contrast, color adjustments, channels, on-screen programming, etc. Using the TV's remote control will let you adjust anything from the comfort of your chair. Complexity of programming and control functions vary between brands and models.

Remote Control.* The nerve center of your home entertainment system. See Chapter 2 for complete remote control information.

Remote Control Finder. Hit a button on the TV and the remote will beep, revealing its location.

StarSight. A subscriber service that sends your TV a complete listings of the week's broadcast fare. Great for simple VCR programming if you're willing to pay the $5-per-month fee.

User Programmable Audio/Video Presettings.* Lets you *customize* a set of features for each viewer's preferences (such as color, sharpness, contrast, etc.).

DIRECT BROADCAST SATELLITES (DBS)

10-Event Programmable Timer. Turns on the integrated receiver/decoder (IRD) when a specific program is on, and turns it off when done.

16x9 Capable. Means the DBS box can send a 16x9 widescreen picture to the television.

Aluminum Dish. Lightweight and rustproof satellite receiver dish.

DSS Compatible Dish. Capable of receiving DirecTV and USSB programming.

Dual-Output LNB. Feeds signals to two separate IRDs.

On-Dish LED Signal Finder. Some dishes, such as Sony's, come with an attached LED indicator to help easily locate the satellite signal.

On-Screen Menus and Programming Guides.* Each DBS system comes with its own form of on-screen programming features and guide.

One-Button Record. Lets you highlight a program to record from the on-screen guide.

Two Sets of Video Outputs.* Lets you hook up both a VCR and a TV to the DBS dish.

VIDEO CASSETTE RECORDERS (VCRs)

181-Channel Tuner.* A VCR is able to receive 181 channels. Be aware, however, that this does not mean the VCR is cable-ready.

Audio Dubbing. Lets you *add* narration or music onto a tape even after you have recorded something onto it. A great editing feature!

Audio/Video Jacks.* All manufacturers provide two to three separate sets of jacks on VCRs. One set includes the standard 75-ohms RF input and output jacks, and another consists of audio/video input and output ports. Additional jacks now offered are the S-input or S-video jacks. These separate the luminance and chrominance signals for better recording and playback.

Auto Clock Setting.* Sets the time on the VCR by receiving a special signal from the local PBS station. See *Plug-and-Play* below.

Auto-Head Cleaning. When you load or unload a tape, the VCR sweeps the heads with an internal pad to remove debris.

Auto Turn-On/Rewind/Shutoff/Eject.* The VCR will turn itself on and play when you insert a tape. When a tape ends, it will rewind itself; then the VCR will shut off and eject the tape.

Cable-Ready.* Receives cable channels if your cable company doesn't require a special cable decoder.

Childproof Lock. Locks the cassette door and shuts out the controls so children cannot insert "unwanted" objects into the VCR or record over valuable tapes.

Commercial Editor. While recording a TV show, your VCR marks suspected commercials and skips over them.

Digital Auto Tracking. The video goes into a feedback mode to automatically adjust the VCR's tracking.

Fast Motion. This is another advantage to four-head (or more) VCRs: you can scan at a faster speed, and the picture will maintain its quality.

Flying Erase Head. If you will be doing any kind of editing, this feature will allow near-professional-quality transitions between scenes. This feature is usually available on high-end VCRs and camcorders.

Four Video Heads.* A cheap VCR will typically have two video heads, and a mid- to upper-range model will have four video heads. The four-head unit will provide fantastic freeze-frames abilities. Some video heads are "oval cut" to eliminate ghosting and color beats.

Chapter 9: Features

Go-To Feature. Searches the tape automatically for a known scene or for a blank section for recording.

Hi-Fi Stereo.* High-fidelity stereo; see Chapter 4. This is the little gem that sets a diamond-laced VCR apart from a CZ cheapy (mono). Most manufacturers offer this feature for around $50 more than mono.

Index Points Locator, Index Search. While recording, you can set index points for later reference. The VCR uses these to access scenes; similar to the go-to features.

Jog/Shuttle Controls. The large dial-like controls on the front of the VCR or remote. They allow you to control the scanning functions, such as variable slow motion, multispeed search or one frame at a time.

Plug-and-Play, Dual Quick Set.* The ultimate no-thinking-needed feature. Simply plug in you VCR and it automatically sets your clock and programs all of your channels.

Quasi S-VHS Playback. Plays back S-VHS tapes but won't record in S-VHS resolution.

Quick-Start. Allows the VCR to go from *Stop* to *Play* in one second as opposed to the typical five seconds.

Slow-Motion. Advances the tape slowly while retaining a viewable picture.

StarSight. See *TV* features.

Super-High-Speed Rewind. Self-explanatory.

Tape Counter and Indicators. Show the tape's position on an LED or as an on-screen function. Most useful for editing sequences.

Timers. Let you preset the recording times of shows.

Title Generator. Same feature on a camcorder. It superimposes text onto your recordings.

VCR Plus+.* See Chapter 4. Look for a unit that controls your cable box.

DVD

10-Bit Digital-to-Analog Converter. Provides information to improve the transfer from digital to analog, resulting in a better picture and sound.

500 Lines of Resolution. DVD will produce 500 horizontal lines of resolution. Quite stunning, especially on large TV screens.

CD/DVD Optical Pickups. Play regular CDs or DVDs on the same unit.

Digital Video Noise Reduction Control. Removes tiny flecks and specks of unwanted color in the picture.

NOTE: Most DVD units operate with a single-layer, single-sided disk. Future units will have twice the capacity with double-layer disks, and four times the capacity with double-layer, double-sided disks.

Dolby Digital, AC-3/Pro Logic/PCM Stereo. Provides you with a Dolby Digital Surround Sound, Pro Logic or PCM stereo signal to pump into your home theater stereo system.

Letterbox, Pan-and-Scan, 16x9 Capabilities. View a movie in the screen format of your choice.

MPEG-2 Decompression. This video standard packs more video into a smaller space.

S-Video Output. High-quality output jacks for a superb picture.

CAMCORDERS

5-Pin Edit Jack. Offers convenient connection to stand-alone edit controllers or computer interface devices.

AA Battery Capability. Standard AA batteries can be used as a power source in the camcorder.

Auto Iris. Automatically adjusts the lens opening for varying light conditions.

Auto Light. The camcorder will adjust to varying lighting situations according to brightness conditions, for improved picture quality in color and tint when shooting in low light levels.

Auto Pause. Automatically stops recording when the camcorder is pointing straight
down for more than two seconds.

Auto Shutoff. Shuts off the camcorder if it is paused for more than a few minutes. Saves on batteries.

Black Fader and Wipes. A special-effects function that fades the picture and sound to black or produces professional wipes.

Character Generator. Superimposes text onto your movie while filming.

Date/Time Display, Battery-Remaining Display, Tape-Remaining Display and End of Tape.* Functions displayed in the viewfinder. You can also display the time and date, superimposed on a recording.

Digital White Balance. Keeps colors "true" with varying lighting situations, such as low light, blue skies or fluorescent lighting.

Electronic Image Stabilizer.* Takes that earthquake shake out of your movies! Picture stabilizer differentiates between intentional and unintentional camera-shake, and compensates without affecting deliberate pans and tilts.

External Microphone Jack. Lets you plug in a microphone to move it away from the noisy cassette components.

JLIP, Joint Level Interface Protocol. A terminal which allows the interconnection of A/V equipment, such as your VCR and a computer.

Motion Sensor. With this function, the camcorder starts recording 3 seconds after it perceives motion, and stops 30 seconds after motion ceases.

Negative/Positive Transpose. Changes the scene to a negative video image. By placing a photo negative in front of the camcorder lens, you can see and record the positive photo image.

R.A. Edit (Random Assemble Editing) Function. Can be operated via remote control. This enables the programmed rearrangements of up to eight scenes at a time, with selections of up to five digital effects and seventeen different scene transitions.

Remote Control.* Most newer camcorders come with an infrared remote.

Self-Timer. Delays recording for about 10 seconds.

Super Lo-Lux. * If you film scenes that are low-lit, choose a camcorder with a low-lux rating; also called Super Lo-Lux.

Titles. You can type short text messages to superimpose onto the tape.

Variable-Speed Zoom. Most high-end camcorders come with this feature. It allows you to zoom in or out of a scene at any speed you wish.

Wide Mode. Some camcorders can simulate a letterbox-type format (16:9).

See the appendices for company Websites and other information that help you find information on features as they become available.

CHAPTER 10

BASIC BUILDING BLOCKS OF VIDEO

> *"Inside every digital circuit, there's an analog signal screaming to get out."* Al Kovalick, HP

Video hardware and accessories are similar to Lego blocks: they both contain basic building blocks to help you assemble your toys. In video, we are dealing with cables, connectors, matching transformers, band splitters, signal splitters, A/B switches, wall outlets, antennas, amplifiers, video selectors and, of course, the equipment itself. Here's we will explore the purpose of these hookup devices and how to use them.

Don't be afraid of hooking up a system. Once you have learned the basic building blocks, you will be able to figure out just about any weird video configuration dangling about.

HOOKING UP YOUR HOME ENTERTAINMENT SYSTEM

The piles of unseen wires in the back of a home entertainment center rear their ugly tentacles and attack when least expected. You may find yourself saying, "I swear there was only one plug on this VCR when I left the store. Where did all these wires and connectors sprout from?" More importantly, you should ask yourself where these wires and connectors should go.

Home entertainment systems require the various video, audio, RF and control circuits to be interconnected, so information can be sent from one component to another, to another. At this time, most systems require a physical connection through wiring. The wiring includes the miles of copper lines and connectors most of us dread encountering.

Not to fear. This section will give you the needed knowledge to conquer those hideous, wiry creatures attached to your components. Attack!

CONNECTOR AND CABLE BASICS

THE PURPOSE OF WIRING

Electricity needs a conductive path through which to route power and signals. Unfortunately, air is not a good conductor; but not many substances are. Copper, steel and aluminum, however, do make good conductors for our purposes. This is what makes up most of our entertainment system wiring.

Wire localizes power or a signal and directs it over a distance. That distance may be miles or, in most video cases, feet. A connector is attached from the transmitting piece of equipment to the wire. A similar connector is attached from the receiving piece of equipment to the same wire. In this way, the signals and power are routed throughout your entire home entertainment package.

CONNECTORS

A connector can be anything from a wall plug to a complex SCART connector. Every signal wire needs a physical connection. Some connectors are in fact multiple connectors in one package.

A connector typically has a female and male component. See Figure 10-1. Patch cords and cables are usually terminated with a male connector, while the female counterpart is on the equipment itself.

A cable is the actual path the signal or power takes to your equipment. When dealing with video and audio applications, the cables require special insulation and shielding to prevent signal loss or interference. They are usually composed of one or more signal wires wrapped in various layers of metal and foam or other insulation. Always insure that you are using these shielded cables properly.

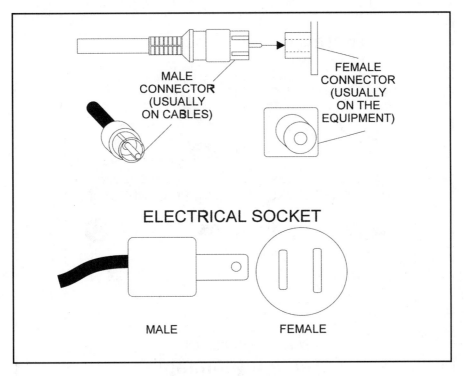

Figure 10-1. Male and female connectors.

CONNECTOR AND CABLE TYPES

See Figure 10-2 and Figure 10-3.

RF Connectors. *RF* means *radio frequency*. There are two basic types of RF connectors:

F-Connectors for Coaxial: F-connectors are used on the ends of 75-ohm coaxial. The connector screws onto the female connection on the back of the equipment.

There are two basic F-type connectors. One is a screw-on male, and the other is a push-on male. A screw-on male is good for equipment that has a permanent placement. It is relatively strong and resistant to signal loss and noise if connected properly. See *INSTALLING F-CONNECTORS ONTO COAX* for more information.

The push-on male connector is great for equipment that is continually moved, such as a well-traveled VCR. The problem is that the connec-

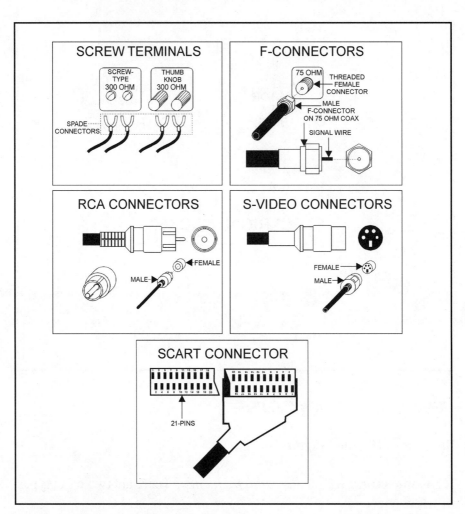

Figure 10-2. Common connector types used in home entertainment systems.

tors wear out in a short period of time, causing signal loss and noise from poor connections.

HINT: Try to use well assembled screw-on F-connectors whenever possible to prevent signal loss and noise from haphazard connections.

Screw Terminals for Flat-Lead Cable: 300-ohm cable is terminated with a bare wire or preferably a spade connector. On the component side there is a screw-type terminal, to which you would either wrap the wire or connect the spade. There are two basic kinds of screw terminals: an actual screw, and a thumb-knob connector.

RF Cables. *Flat-lead (Twin-Lead)*: This is the flat, wide cable used in older antennas and video equipment. It is commonly used for VHF and UHF signal routing. The impedance on this type of cable is 300 ohms,

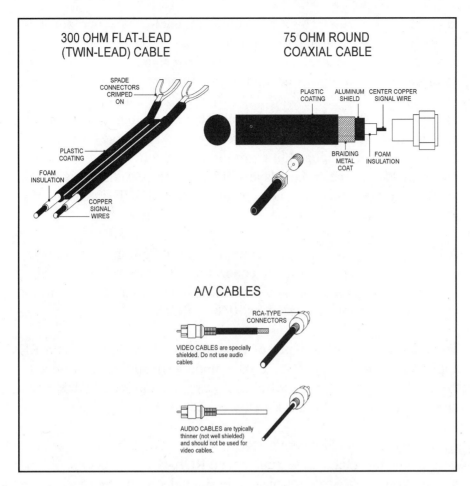

Figure 10-3. Common cables used in home entertainment systems.

and thus not directly compatible with 75-ohm coaxial cable: a matching transformer is required for the joining of the two.

If you are going to use this relatively cheap substitute for coaxial cable, be aware that it weathers and cracks easily, is susceptible to interference from other signals, loses signal strength when split to several components, and is prone to AC interference. However, if you need to route UHF signals around your home theater, it is an acceptable cable. It is often used because it offers less signal weakening and better UHF signal transfer.

If you are installing a new antenna or routing RF signals around to the various components, I highly recommend using shielded coax.

Chapter 10: Basic Building Blocks of Video

Round Coaxial Cable: This is the familiar rounded cable (approximately 1/4" in diameter) used in most modern RF video signal applications. It is used to route the antenna's VHF and UHF signal to the TV, and to connect the various component RF signals together.

This is by far the most popular wire to thread to your equipment. It provides excellent guard against interference. The cable consists of a center signal wire wrapped in foam insulation, foil, a solid or braided metal shield, and one or more layers of insulation. Coaxial is also more durable than flat-lead cable and transmits VHF and UHF signals equally well.

HINT: Coaxial cable has had many modern improvements, making it the best all-around choice for video applications.

All TV-type coaxial cable has an impedance of 75 ohms, and thus requires a matching transformer to connect to 300-ohm cable or 300-ohm RF connections. Make sure you don't mistakenly purchase cables meant for CB and amateur radio, such as RG8 or RG58.

There are two common types of coaxial cable used in television RF applications: RG59 and RG6. For common VHF applications, go with the cheaper RG59. For high-frequency applications such as UHF and direct broadcast satellites, it is better to use the RG6.

Audio and Video Connectors (RCA-Type). You can recognize A/V lines by the fact that they terminate with RCA-type connectors. An A/V patch cord is used to route the composite video signal and audio (one line for mono, a left and right line for stereo) to the various A/V components. A/V lines are a much more efficient method of connecting A/V devices as opposed to RF cables; the signals stay neat and separated.

Video-Quality Shielded Cables. When you need to connect a composite video *IN* to a composite video *OUT* (or vice versa), always use a special video-quality shielded cable. DO NOT USE AUDIO CABLES IN PLACE OF VIDEO CABLES. The impedance is incorrect and there is insufficient shielding.

The video cable is a high-quality shielded cable with RCA-type male connectors at each end. They are usually 75-ohm cables, and can resist much more interference than a cheap audio cable. If you can afford it, try to purchase the gold-plated connectors for maximum performance.

> **INSTALLING F-CONNECTORS ONTO COAXIAL CABLE**
>
> Special tools are not needed to install F-connectors onto coaxial cable. All you need is a combination wire stripper/crimping tool. Have patience while installing them. A good, clean RF signal is the result of strong proper connections. Here is the proper method of doing this:
> 1. Cut about 1/2" of the outer plastic insulation jacket off with coaxial or wire strippers.
> 2. Fan the metal braiding back and cut it to a 1/8" length all the way around.
> 3. Peel and tear off the aluminum shielding and strip away the foam inner insulation to expose about 1/4" of the shiny copper signal wire.
> 4. Place the F-connector over the cable and crimp it with pliers or a crimping tool. Most wire cutters have this crimping tool built into them.

Audio Cables. Audio cables are also terminated with RCA-type connectors, but are not as well-shielded as video cables. They are typically thinner and contain a matched set of wires, left and right, for stereo signals. Try to use high-quality cables, and if possible, get the gold-plated RCA connectors.

S-Video Connectors and Cables. S-inputs and outputs are used to split a demodulated video signal into its separate components: luminance and chrominance. This allows for the transfer of a crystal clear picture between such components as VCRs, TVs, camcorders, DBS, DVD, etc. The connector is round with four circular pins and one square pin inside. If your video equipment supports this connection, use it whenever possible.

SCART Connectors and Cables (A/V Euroconnector). This is a special cable used in some video equipment that attaches audio and video signals. It also transfers data and control signals. This allows one piece of equipment to control several others, and also allows for a better separation of signals for a high-quality sound and picture. SCART is used as a trouble-free connection for all current (and future) A/V equipment, such as high-end television, laser disks and digital video disk players and VCRs.

VIDEO AND AUDIO CONNECTIONS EXPLAINED

Each piece of equipment has video input and output connections. Here is a brief explanation of each to help you better understand hookup methods:

VIDEO
Component Video, Best Method. Some video components come with a special component video connection. It uses 3 RCA connectors and cables: one for the red video signal, one for the green signal, and the last for the blue signal. A SCART connector has these lines built into it. Most new high-end video applications such as DVD have this feature, to deliver pristine color images.

S-Video, Good Method. Most new video equipment has these connections. They slip the video into its luminance and chrominance signal for a quality transfer.

Composite Video, OK Method. This is the old RCA cinch (baseband) standby. This is a shielded video cable with RCA-type connectors on the ends. It contains the combined luminance, chrominance and video timing signals on one line.

RF Combined Signal, Terrible. This is the line and connectors that carry the combined video and audio baseband signals. Try to avoid their use if possible, as the continual modulating and demodulating causes signal loss.

AUDIO
Digital Audio, Best Method. New video equipment comes with a special digital audio connection. This carries the Dolby Digital Surround sound signal through a 75-ohm cable or a new high-speed fiber optic connection.

Component Analog Audio, Good Method. Some new equipment provides an analog path for the Dolby Digital signal. It uses 6 audio RCA cables or a special DB-25 pin connector similar to a printer cable.

Stereo and Surround Analog Audio, OK Method. This is the most common baseband audio link. It uses 2 RCA lines to carry a stereo signal or a Dolby Pro Logic, or standard surround sound signal.

The SCART connector has 21 pins and is rectangular shaped. The cable is quite thick compared to other video cables because of the number of wires involved.

NOTES ON CABLES AND CONNECTORS

Here are a few things to remember while hooking up cables to your equipment:

Impedance. This is the measurement of the opposition a signal encounters while traveling through the cable. It is measured in ohms. In video applications, we are dealing with two difference impedance values: 300 ohms (flat-lead) and 75 ohms (coaxial cable). If you see these numbers on the back of a piece of equipment, it means to only attach those value cables to the appropriate outlet or inlet.

Optimizing Video Signals in Your System. RF lines carry a modulated (combined A/V) signal that is demodulated (uncombined) when it enters a component, and recombined for transfer to the next component. An example would be an RF cable coming into a VCR then going back out to a TV.

If you use RF lines, the combining and uncombining of a signal in each component will cause signal degradation as the signal goes down the line, resulting in a poor picture and sound. Therefore, it's better to route signals around your entertainment system with the A/V, S-Video or SCART lines whenever possible. Continuing the with the above example, this means you would use A/V lines instead of an RF cable to transmit the signals to your TV.

Quality of Cable. Cable is measured by how much signal is lost for a given length of wire in a given frequency range. For example, a 3-dB signal loss per 100 feet on VHF channels is better than a 5-dB signal loss per 100 feet. Three decibels may not seem like a lot for 100 feet, but if you are mounting an antenna quite a ways away from the TV, the lengths will stack up against you.

Whenever you add to a home entertainment system, remember that the higher the quality of the cables and connectors used, the higher the quality of the sound and picture.

OTHER BASIC BUILDING BLOCKS OF VIDEO SYSTEMS

Here are some other basic components of a home entertainment center:

Signal Amplifiers. When a signal is weak or has traveled through a long cable, it may need an amplifier connected in-line. Antennas usually have an amplifier somewhere in their loop, to boost the signals coming from various TV stations. See the HOOKING UP AN ANTENNA section for more information on signal amplifiers.

Matching Transformers. Because flat-lead cable is 300 ohms, it is not directly compatible with 75-ohm coaxial. You need to put a *matching transformer* in-line with the two. See Figure 10-4. There are two types of these transformers. If you are connecting a twin-lead line to a your TV's coaxial female connector, you need a 300- to 75-ohm unit. If you are connecting a coaxial cable to a two-terminal connection, you need a 75- to 300-ohm unit.

Band Splitters. If you want to split the VHF and UHF signals into separate lines, you need a *band splitter*. For example, a coaxial is coming in from a VHF/UHF antenna. You should attach a band splitter from the antenna to the TV: the input goes to the antenna, the output VHF coaxial goes to the TV set, and the UHF line goes to the screw

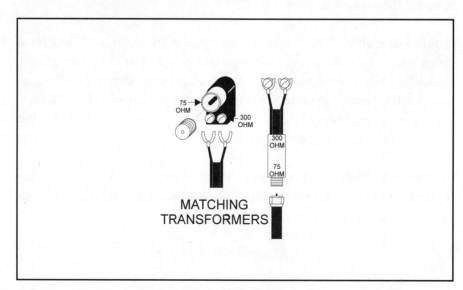

Figure 10-4. Basic building blocks of video: matching transformers.

terminals. Most new televisions have an internal band splitter, eliminating the need for an external wiring mess.

Signal Splitters. If you are going to split an RF signal, you need a special *signal splitter*. For example, if you have a coax coming into the home for cable TV and want to split it for use on multiple sets, you should connect the main line to one side and the various TV antenna lines to the remaining connectors.

You can split the input signal into two, three or even four separate outputs. Be aware that each separate output line will be half, a third, or a fourth of the full signal strength. If the separate signals are too weak, use a distribution amplifier, which will strengthen each signal before it exits the unit. Be sure to cap off each unused output with a special "terminator." This prevents signal leakage and possible ghosting problems.

High-Isolation A/B(/C) Switches. If you want to switch between two or three different signal sources, you can install a *high-isolation A/B or A/B/C switch*. With the flip of a switch, you can change between cable and antenna, or cable and DBS, etc. A 75-ohm cable is the output signal, and the input is through two or three 75-ohm connectors.

Outlets. When a cable is installed, the cable company usually places the 75-ohm cable hookup into a wall outlet very similar to an electrical wall outlet.

Also, some DBS services require that the integrated receiver/decoder be hooked up to a phone jack. An Internet appliance typically has a modem that needs to be connected to a phone line as well.

Video Selectors. When you have to connect tons of equipment at once, it's a good idea to have a video selector. This is a box with a series of input and outputs in the rear and some type of selector interface on the front. It lets you *select* between various signal sources and outputs. For example, if you have two TVs, a cable box, and two VCRs in one system and wanted to select the source to display, you'd simply click the appropriate buttons on the video selector.

In some cases, this little black box replaces masses of splitters and A/B switches. It is also a more convenient way to change signals as you simply mount the box atop the television for easy access.

Equipment. Of course, the most important building block in any home entertainment system is the equipment itself. This includes televisions, VCRs, antennas, cable decoders, integrated receiver/decoders, satellite dishes, laser disk players, digital video disk players, video selectors, camcorders, editing equipment, and any new equipment that comes along. Each piece of equipment usually has the controls in front and the input/outputs in the back. Video equipment uses various accessories, already described, in conjunction with cables and wires. To put a system together requires a working knowledge of each component; so if you are still unsure of what each unit performs, restudy the previous chapters or research them further so you can install them with confidence and less hassle.

Equipment Inputs and Outputs. Your home entertainment center is a series of INs and OUTs. Each piece of equipment's inputs and outputs are chained together to produce sights and sounds. For example, a coaxial cable is affixed to a cable box's IN connection. The cable box OUT connection has a cable running to the TV's IN connection. Later in this chapter we will describe the typical patterns of INs and OUTs used in a common system.

PLANNING YOUR CONNECTIONS

Placement of Components. It is important to have the placement of the components in mind, whether it be on an existing shelf, a simple TV stand, or a full-blown home entertainment center cabinet. In fact, some people even prefer to build the unit themselves and meld it into the decor.

Once you have a place for the home entertainment system in mind, plan out where the television will sit, where the VCR will reside, etc. If you are placing them in a cabinet, is there enough room? Do you need to run special long cables? Will there be enough air circulation to stop the components from prematurely burning out? Planning each step ahead of time will save you heartache, time and (most importantly) having to replace video equipment every other year.

After you have the component rack and television location all plotted out, think of the speaker placements. If you have a surround sound system, where do you want the surround speakers to be? See Chapter 2, *Surround Sound Speaker Placement*, for further planning tips. Is there enough room for this all-out sound system, is a preliminary question to ask. Others questions to be considered: Do you have enough special audio cable to run to all the speakers? Can you build the speakers into the wall? Do you need to hide the wires? Where should you plant the subwoofer?

NOTE: You should be able to put a subwoofer anywhere in a room. Humans cannot determine the direction of a low-frequency signal, so subwoofer placement can be next to the TV, behind a sofa, or even in a hidden corner of the room.

BASIC CONNECTION RULES

Here are a few things to keep in mind while installing components:
- Always connect an IN to an OUT, or vice versa. Never connect an IN to an IN or an OUT to an OUT.
- Never hot-wire a 75-ohm line to a 300-ohm connection (or vice versa) without a matching transformer.
- Use the highest quality cable, connectors and accessories possible to ensure proper picture and sound.
- Try to place the antenna as close to the TV as possible because of the above.
- Look for RG59 coaxial cable or RG6. Preferably, use RU-6U. Other forms of coaxial are useless for TV applications as they are intended for other, non-TV signal transmissions.
- Study the equipment's manuals to make sure you have the equipment to handle any hookup procedures that are out of the ordinary.
- Know what each connection is actually for and where it leads.
- Know the basic building blocks inside and out.

KNOW YOUR BASIC BUILDING BLOCKS OF VIDEO

By thoroughly understanding these basics, you will be able to configure just about any system to meet your needs. The following explanations are guidelines to further help you learn how to hook up various components. Remember, these guidelines are not set in stone, and should simply act as a tutor to help you learn the basics.

HOOKING UP AN ANTENNA

Hopefully, after reading Chapter 3, you know what type of antenna is going to roost on your roof. See Figure 10-5. Now it is time to examine the accessories you will need for it, as well as how to piece it together.

Mountings for Antenna. In order to mount an antenna, you need a mounting kit. This would include various clamps or specially-designed mounts. You have the option of mounting the antenna on a chimney, on a vent pipe on the roof, on the peak of the roof, on a mast, on a DBS dish, etc. Whichever you choose, make sure you purchase the proper mounting hardware kit.

Signal Amplifiers. If you are far from a TV broadcast tower or are running a lengthy cable, you need to install a signal amplifier with the antenna. There are several types that provide a specific need. See Figure 10-6.

Amplifiers are rated in *signal gain* (decibels - dBs). The best signal-to-noise ratio means the best picture possible. Always remember: amplifying a signal that is already weak (snow on the TV) due to a poor

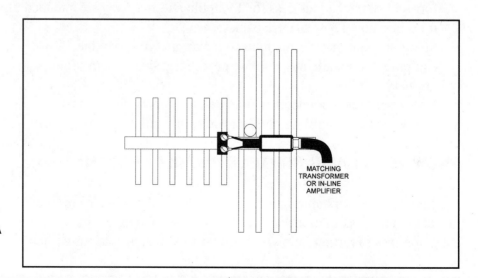

Figure 10-5. A typical rooftop antenna.

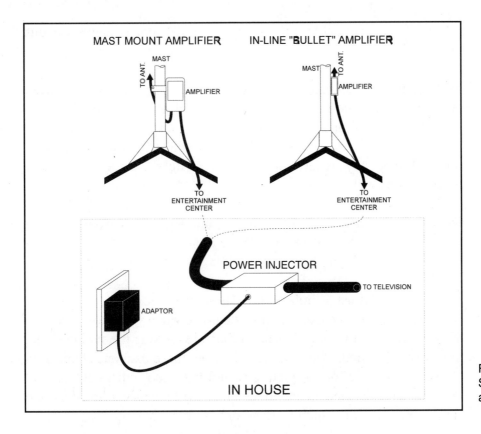

Figure 10-6. Signal amplifiers.

antenna choice is useless; you will only make the poor signal worse (by amplifying the snow)! Use a better antenna with a higher gain instead.

Mast Mount: This is a signal amplifier that is attached to the mast of the antenna. It puts the amplifier as close as possible to the signal from the antenna itself. This means that a signal will not have to travel down a lengthy cable and be amplified after losing most of its strength. Do not use a mast mount amplifier to boost a weak signal from a lengthy (100+ ft.) cable. Instead, use it to boost an originally weak antenna signal.

If you are only using the mast mount amplifier to boost a VHF or UHF signal, a 20- to 25-dB unit will work fine.

In-Line "Bullet" Amplifiers: These amplifiers are used in-line with the coaxial cable itself. It is best to use them as close to the antenna as possible. Their main function is to boost the signal just enough to compensate for a long cable run. A 10- to 12-dB unit is plenty for this purpose.

Both the mast mount and in-line amplifiers use a special device placed at the location of the entertainment center to power their innards. They "inject the power" up the coaxial and into the amplifier placed near the antenna. Make sure the power has a straight line to amplifier and is not routed through any signal splitters or A/B switches: these devices may block the power.

Distribution Amplifiers: These are similar to signal splitters except that they amplify the output signals. These are best to use if you are splitting a cable or antenna signal to multiple TV sets in the house.

Attenuators, Ghost Eliminators. These are used when you want to make a signal weaker. An example would be when you are too close to a TV station and the antenna's amplifiers are causing the station signal to be overamplified. This causes you to see more than one channel on the screen at once, or the picture would have too much contrast and become dark. The attenuator fixes this by allowing you to adjust the signal strength reduction by a fixed or variable amount.

The attenuator can also be used to eliminate ghosting. A ghost eliminator is merely a variable attenuator.

Band Splitters and Combiners. The band splitter is the same as described earlier in this chapter. A band combiner is the opposite; it combines the VHF and UHF signal to send to a signal cable. If you are using separate VHF and UHF antennas, use one of these to send the signal to your video equipment.

Filters. If you are receiving picture interference from a CB radio, shortwave radio or FM station, you need to add a filter to remove it. The interference is best diagnosed by these symptoms: snow, static, streaks in the picture, etc.

Another type of interference that causes similar difficulties is from the AC line. This can be caused by vacuum cleaners, microwaves, hair

SETTING UP A SURROUND-SOUND STEREO SYSTEM

No home theater is complete without a rock-the-house surround sound system. You have bought the Dolby Pro-Logic receiver and speakers, and are now stuck. There is more to putting together a surround sound system than just hooking up wires and cables. You have to make sure the speakers are set up properly.

First determine the placement of the television and your main viewing seat.

Now plan where the speakers will be planted:

Left and Right Speakers. These need to be placed on both sides of the television and angled toward an imaginary center point, where the viewer sits. Make sure the speakers are aligned flush with the television's face. Draw an imaginary line from the viewer's position directly to the center of the television's screen. The left speaker should sit about 22.5 degrees to the left of that line, and the right speaker should sit 22.5 degrees to the right.

Center Dialog Speaker. This is placed atop the television set or below it on a rack. It should be placed equal to or slightly behind the same plane as the left and right speakers, but never ahead of them. Make sure the mid- and high-frequency drivers of the three front speakers (left, right and center) are as close as possible to the same height. Make sure the center speaker has magnetic shielding to prevent damage to the TV tube.

Surround Speakers. They should be placed on either side of the viewers, and two to three feet above them. Angle them at each other, not pointing down to the viewer. If this placement is not possible, you may want to experiment by placing the speakers behind the couch and bouncing the sound off a wall back at the viewer. If the sound envelops you it is fine. If it sounds like it is coming from behind you, changed the configuration.

A Subwoofer can be placed anywhere in the room.

Once you have the speakers in place, THEN it is time to start thinking about running the wires. Keep in mind that the speakers will require more and longer wire than any of your other components. Can you hide them along a baseboard, or do you want a professional install with the unsightly wires hidden in the walls?

Once you are all set up, and the receiver/amplifier is up to speed, give it a test. Dolby Pro Logic decoders feature a test signal generator for balancing channels. A sound field will rotate clockwise around you. Adjust the balance controls until each channel plays at the same level. Time to loosen the roof's plaster with a movie soundtrack!

dryers or most motor-operated items. If you are having problems with this, use an AC line interference filter between the equipment's power supply and the wall socket.

HOOKUP DIAGRAMS

The following section provides diagrams to help you familiarize yourself with common video setups. This is by no means a complete directory, but it will provide you with a basic tutorial in hookups.

RF AUDIO

See *VIDEO* above.

HOOKING UP A TELEVISION WITH A BAND SPLITTER

Figure 10-7. If you have a combination VHF/UHF antenna, you will need to use a *band splitter* to divide the signals once they reach the television.

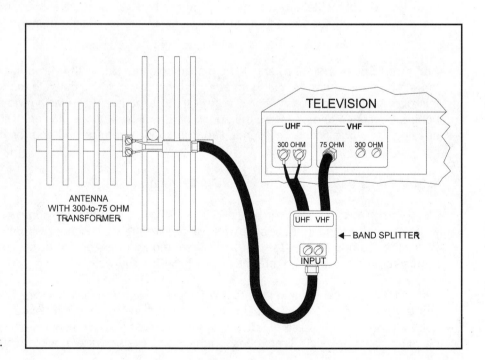

Figure 10-7. Hooking up a television with a band splitter.

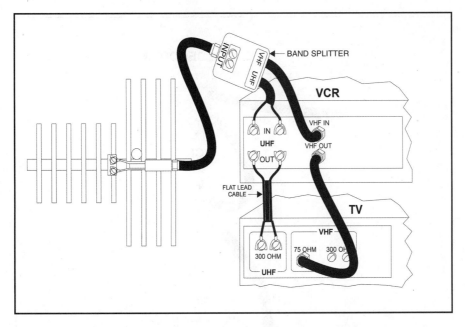

Figure 10-8. Hooking up a television and VCR with a band splitter.

HOOKING UP A TELEVISION AND VCR WITH A BAND SPLITTER

Figure 10-8. If you are hooking up a VCR between the antenna and the TV, hook a band splitter to the VCR. Run a 75-ohm coaxial from the VHF *OUT* on the VCR to the VHF *IN* on the TV. Also, use a 300-ohm twin-lead cable for the UHF signals between the VCR and TV.

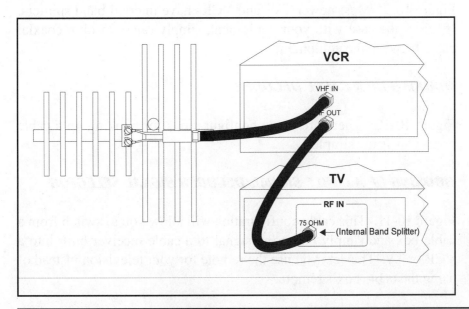

Figure 10-9. Hooking up a television and VCR with an internal band splitter.

Chapter 10: Basic Building Blocks of Video

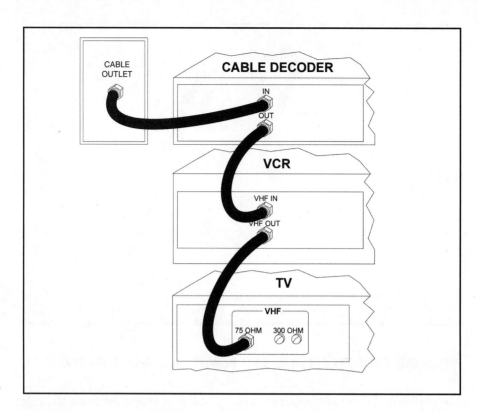

Figure 10-10. Hooking up a cable decoder.

HOOKING UP A TELEVISION AND VCR WITH AN INTERNAL BAND SPLITTER

Figure 10-9. Most newer TVs and VCRs have internal band splitters. If this is the case with your equipment, simply run a 75-ohm coaxial cable between the equipment.

HOOKING UP A CABLE DECODER

Figure 10-10. There are many configurations for hooking up a cable box. This is the simplest.

HOOKING UP A CABLE SIGNAL/DECODER SIGNAL SELECTOR

Figure 10-11. This cable configuration will allow you to switch from a cable box and simply pump the signal to a cable receiver built into a VCR or TV. This lets you use the remote for your television instead of using the cable box's remote.

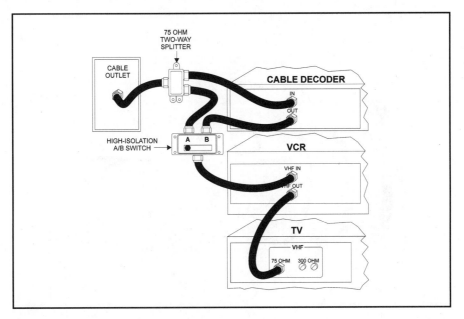

Figure 10-11. Hooking up a cable signal/decoder signal selector.

HOOKING UP A CABLE/ANTENNA SIGNAL SELECTOR

Figure 10-12. When the cable signal goes on the fritz (which can happen quite often), you can still rely on an antenna. By hooking up an A/B switch, you can change signal reception from the cable to an antenna with ease. Sometimes the antenna signal may even be an improvement on a cable signal.

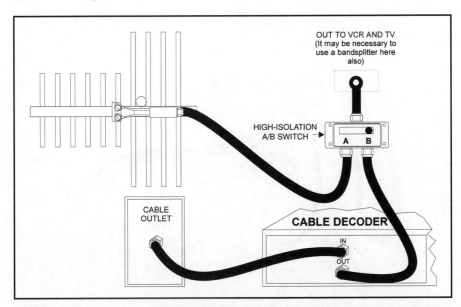

Figure 10-12. Hooking up a cable/antenna signal selector.

Chapter 10: Basic Building Blocks of Video

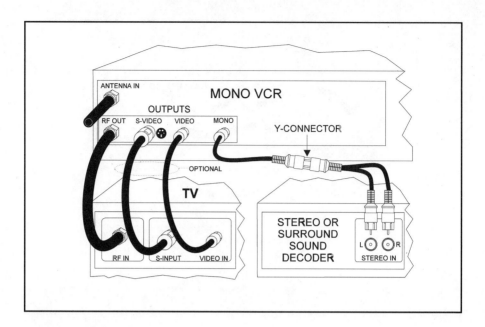

Figure 10-13. Hooking up a mono VCR.

HOOKING UP A MONO VCR

Figure 10-13. If you have a mono VCR and want to hook it to a stereo receiver/amplifier, use this configuration. Simply run an audio cinch cord from the VCR's audio port to a Y-connector, then connect the other two ends to your stereo's left and right input channels.

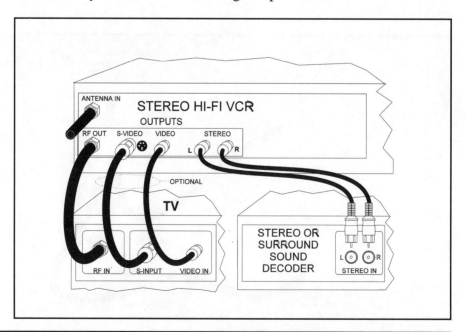

Figure 10-14. Hooking up a stereo hi-fi VCR.

HOOKING UP A STEREO HI-FI VCR

Figure 10-14. Hooking up a stereo VCR is as simple as connecting two audio patch cords between the VCR and the stereo receiver/amplifier. If your VCR and TV have S-video connectors, use them to transmit the video signal. Otherwise, use RCA video lines.

HOOKING UP TWO VCRs TO COPY TAPES

Figure 10-15. There are a few ways to hook up two VCRs for copying video tapes. However, the best method is to use an A/B switch so you can view either VCR's output.

HOOKING UP A DIRECT BROADCAST SATELLITE: ONE LNB/IRD

Figure 10-16. A direct broadcast satellite (DBS) requires an antenna to receive local channels. Here is the configuration to hook up one integrated receiver/recoder to one LNB.

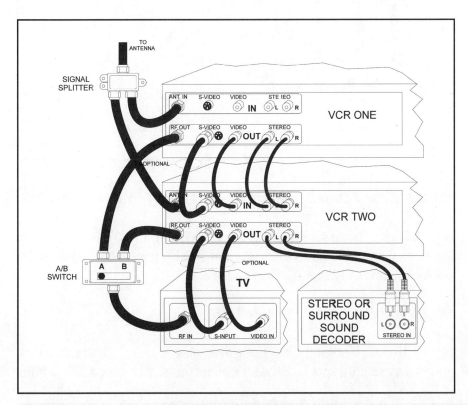

Figure 10-15. Hooking up two VCRs to copy tapes.

Chapter 10: Basic Building Blocks of Video

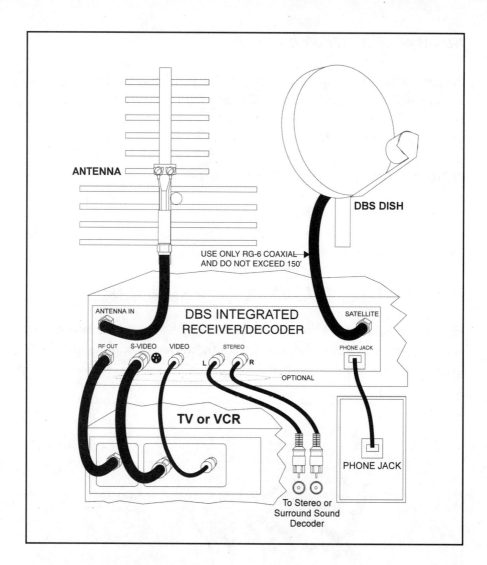

Figure 10-16. Hooking up a direct broadcast satellite: one LNB/IRD.

HOOKING UP A DIRECT BROADCAST SATELLITE: TWO LNBs/IRDs

Figure 10-17. Some satellites have two LNBs built into the dish, so you can hook up two IR/Ds to the same dish. Otherwise, additional televisions can view only one station at a time.

HOOKING UP A LASER DISK PLAYER

Figure 10-18. Hooking up a laser disk player or a digital video disk player may require special SCART connections. Some new DVD units

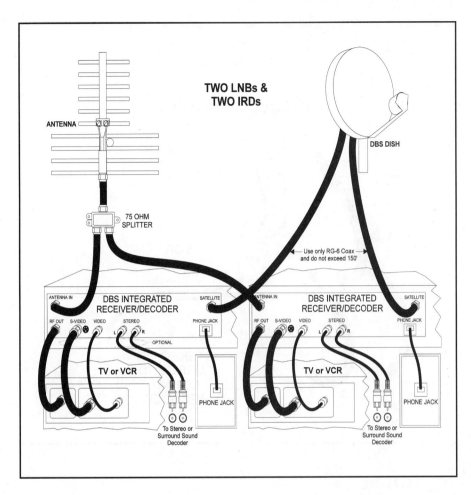

Figure 10-17. Hooking up a direct broadcast satellite: two LNBs and two IRDs.

even require special coaxial-type connections or fiber optics for digital sound.

NEAT HOOKUP GADGETS

There are wireless signal devices on the market now that allow you to receive a signal from a distance. This is great if you want to send a cable or DBS signal to another room in the house. Elcon Technologies also makes a product that uses your AC lines to transmit TV, audio and computer signals throughout the house. It is called ezTV, ezAUDIO, etc. You can find out more information about this at http://www.elcomtech.com/

Chapter 10: Basic Building Blocks of Video

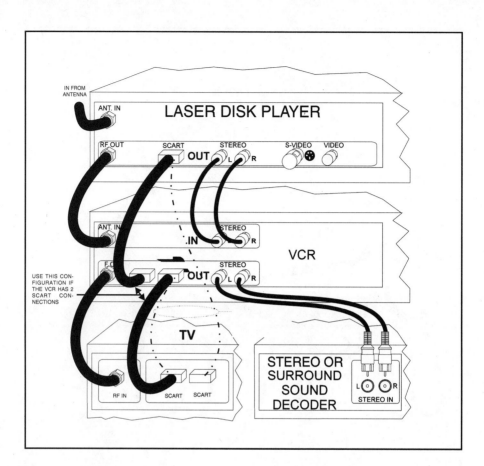

Figure 10-18. Hooking up a laser disk player.

THE FUTURE

Special fiber optic cables will soon transmit all video and audio information digitally throughout your home entertainment system. They will allow mass amounts of information to be transferred at light speed. Maybe this will further complicate hooking up systems; who knows? Maybe wireless components are the answer. Imagine never having to run another wire in your life! In the meantime, learn the basics and keep in mind the end purpose: delivering sights and sounds to your eyes and ears.

CHAPTER 11

TROUBLESHOOTING VIDEO EQUIPMENT

"If you understand it, it's obsolete." Mosaic Computer Inc.

Today, electronic equipment is highly reliable. It runs at much lower temperatures than old vacuum tube technology, and thus lasts longer. However, Murphy's law does apply to video equipment as it does with anything else: if anything can go wrong, it will go wrong. Actually, the video equipment version of Murphy's law is more like, "If the warranty runs out, something will definitely go wrong."

To help ease the pain of realizing that your high-priced audio/visual components could turn into paperweights, I offer these simple-to-follow steps to diagnose and cure whatever ails them:

IT IS USUALLY SOMETHING SIMPLE

Despite what a repair shop says, problems are usually caused by the simplest of things. Many a technician (myself included) has worked for hours to check each circuit only to find out that a switch was not set properly, or some other elementary procedure was not followed. Look for the obvious and simple first:
- A plug not plugged in, or a wall socket turned off by a control switch on the wall.
- A blown fuse.
- A patch cord or other cable going into the wrong connection.
- Thinking the interference indicates there's something wrong with the antenna, when it is actually a hair dryer.
- You think a videocassette is damaged when in fact the VCR only needed to be tracked.

KNOW WHEN TO CALL IN THE PROS

If there is a painful point in every person's life, it is admitting you need help. If you follow the charts in this chapter and still cannot solve the problem, it is time to send the equipment to a repair shop.

Warranty. If the equipment is still under warranty, pack it up and take it back immediately to where you bought it. Don't let the warranty run out.

If it was a factory warranty, you may have to ship it to the manufacturer or take it to an authorized service shop.

Out of Warranty. If you have followed the advice in Chapter 8, then you will probably have some money set aside for the purpose of repairs. Find a reputable/authorized repair shop and get a written quote. Be-

---

THE MICROPROCESSOR'S IMPORTANCE IN VIDEO

The *microprocessor* has advanced video technology by leaps and bounds. Digital technology is finding its way into TVs, VCRs, camcorders, video disk players, and just about every other piece of high-priced equipment in your home theater cabinet. In fact, nearly all analog equipment will soon be replaced by a superior digital counterpart.

Why is the microprocessor so valuable to video? By digitizing an analog signal, the microprocessor is able to manipulate the data in a much more efficient manner. It can clean up unwanted noise, compress the information into a much smaller package, add special effects that used to take stacks of analog equipment to perform, and can more easily interface with computers.

Sets of yesteryear used the ancient power-robbing technology of vacuum tubes. Praises were sung when the solid-state transistor was brought into existence. It meant a reliability factor previously unknown to television. Because of the transistor's low power requirements, it gave off much less heat than vacuum tubes; thus resulting in long-lasting TV and video equipment. Now, with microprocessors controlling each function of a video device's innards, there is a reliability previously unimaginable

The microprocessor also allowed for cheaper, simpler, faster production. Now modern TVs only have a handful of physical circuitry. Cheap televisions with great features now adorn our living rooms. Even the most modest of sets now produce video quality previously available only on top-of-the-line TVs ten years ago. What's next?

cause there are thousands of electronics repair shops in nearly every city, the choice is hard. Try to find one that friends or the manufacturer have recommended to you, and make sure it uses certified technicians. If you were happy with the dealer who sold you the component, you may want to try their service department. Some major electronics stores, such as Radio Shack, even offer repairs for any type or brand of electronics equipment.

Contact a Repair Shop and Get an Estimate. Don't just take your ailing video equipment to a repair shop and leave it there. Give the yellow pages a workout first. Ask the repair shop personnel if they are severely backlogged, if they offer specials such as VCR cleaning deals, and their hourly rate. Don't bother with asking how much it will cost to fix the component: most shops will not answer without looking at it first.

After you have found a shop, take in your component for an estimate. About half the repair shops will offer this service for free. Make sure, however, that it *really* is *free*. Some places will charge you $10 to $50 to reassemble your unit after they take it apart to look at it. If you cannot find a free estimate, never pay over $50.

Repair or Replace. With today's disposable mentality, people are no longer willing to fix a cheap piece of broken equipment. Ask yourself, "Is it worth a $150 repair bill, or should I go down to the electronics shop and shell out $250 for a newer working model?"

Actually, in many cases the repair bill would well exceed a replacement's cost. So before you take your machine to the repair shop, you may want to call around and see how much a new one costs, then compare.

COMMON PROBLEMS

Cable and Connection Problems. Most interference and weak signal problems can be traced to faulty cables or connections. Make sure the connectors are properly applied to the coaxial, that the center signal wire is a shiny copper color and not broken off, and that the cable or wire is plugged into the correct IN or OUT connection. If you are using push-on F-connectors, make sure they are not worn and slipping off the female side.

If you are using RCA cinch cords for video, and are experiencing degraded performance or interference, then you may not be using specially-shielded video cable. You may want to use gold-plated RCA connectors if problems continue.

Miscellaneous Video Equipment and Signal Problems.
- Make sure you have the proper channel for your VCR, video disk player, cable box or DBS IR/D connected to the TV. In other words, if you set the VCR 3/4 switch to Channel 3, then make sure the TV is switched to Channel 3.
- If you are not receiving channels you know are in your area, you may have to manually program them into your auto-programming TV or VCR. When the TV is auto-programming, it may not pick up a slightly weakened signal; you may want to add an amplifier as well.
- Remember that if a DBS system is not receiving a pristine picture, it will not receive anything. So if you have a hazy or fuzzy picture from your dish, then it is because of the lines or equipment.
- Something called Macrovision is used to copy-protect videocassettes. This prevents copies of prerecorded movies from being made. The problem is that it sometimes creates a poor picture or interference when you are playing a tape normally. If this occurs, look into getting a sync stabilizer or a "Macrovision buster."

TROUBLESHOOTING GUIDE

Television

PROBLEM	SYMPTOMS	POSSIBLE CAUSES	POSSIBLE CURES
TV will not turn on.	Power LED is off; TV won't function.	1. No power to TV. 2. Blown breaker or fuse.	1. Plug in the set, switch on the wall switch. 2. Reset breaker or replace internal fuse.
TV comes on but there is no picture.	Power LED is on but no picture. May or may not have sound.	1. Brightness or contrast are turned down too far. 2. VCR or LD is paused. 3. TV circuits faulty.	1. Turn up the brightness and contrast. 2. Turn PAUSE off or turn machine off. 3. Send for service.
Remote control malfunctioning.	Buttons on remote will not operate TV functions.	1. Out of range or not pointing at IR receiver. 2. Remote batteries are dead. 3. Remote is defective or TV circuits in need of repair. 4. Universal remote programmed incorrectly.	1. Move closer to TV or point directly at IR receiver. 2. Replace batteries. 3. Replace remote or take TV in for service. 4. Program remote according to manual.
Picture Quality is poor.	Picture is grainy, snowy or hazy.	1. Channel slightly off tune. 2. Bad cables or connections. 3. Poor signal reception. 4. Signal too weak. 5. Signal is weak from too much loss over a long cable.	1. Fine tune the channel controls. 2. Inspect cables and connections, and replace if necessary. 3. Adjust, repair or replace the antenna. 4. Use a higher gain antenna or amplifier. 5. Install an in-line amplifier if cable is more than 100 ft. long, or a distributive amplifier if the signal is split.
Ghosting	Two images at once appear on the screen.	1. Antenna is so strong it is picking up two TV stations on the same frequency. 2. TV is receiving a reflective signal off a water tower or building. 3. Cable use for antenna is acting as an antenna itself.	1. Redirect antenna or add a ghost eliminator. 2. Same. 3. Use high-quality shielded cable (coax) and keep it as short as possible.
Sputtering interference.	Flecks appear randomly on the screen, and sound is crackly.	1. AC line interference. 2. A close CB or other RF source is causing interference.	1. Add a power line filter. 2. Install an RF filter.

Wavy line interference	Picture is clear but wavy lines appear and sound is affected.	1. RF or CB interference. 2. More than one signal from video devices are mixing with main signal. 3. Cable decoder defective. 4. Signal too strong. 5. Faulty TV.	1. Install RF filter. 2. Turn other device off or service it. Replace RF switch of A/B switch. 3. Service decoder. 4. Add an attenuator to reduce the offending signal. 5. Service TV.
Ground loop interference.	One or two dark bars float up and down the screen.	1. If it is an older set without two different sized prongs, the plug may be flipped. 2. Cable system or antenna not properly grounded.	1. Flip plug or attach all your equipment to a power bar. 2. Ground them properly.
Multiple audio signals	You can hear the audio from another station on the channel being watched.	1. Excessive channel interference from close channel. 2. Interference from CB or other RF source. 3. Cable decoder defective. 4. Signal too strong.	1. Fine tune signal. 2. Install RF filter. 3. Service decoder. 4. Add an attenuator to reduce the offending signal.

Satellite Dish

PROBLEM	SYMPTOMS	POSSIBLE CAUSES	POSSIBLE CURES
IR/D won't turn on.	Power LED off and no controls.	1. No power. 2. Tripped breaker or blown fuse.	1. Plug in unit or switch on the wall switch. 2. Reset breaker or replace internal fuse.
IR/D turns on but won't function.	Power LED on but no control functions.	1. Circuits defective.	1. Service.
Can't receive.	Will not receive any channels.	1. Card is not inserted. 2. Dish is not correctly aimed. 3. You have not paid to watch that particular channel.	1. Insert card. 2. Aim dish properly. 3. Order pay-per-view.
Can't receive local station.	Picks up DBS channels but not local channels.	1. No antenna is attached. Remember, DBS doesn't receive local channels without an antenna.	1. Attach antenna to IR/D.
Picture fuzzy.	Picture comes in with interference.	1. A digital signal comes in pristine or not at all. The problem is somewhere in the cable.	1. Repair cable or add amplifier.

VCR

PROBLEM	SYMPTOMS	POSSIBLE CAUSES	POSSIBLE CURES
VCR will not turn on.	Power LED off; nothing will work.	1. No power. 2. Tripped breaker or blown fuse. 3. Timer is on.	1. Plug in the VCR or switch on the wall switch. 2. Reset breaker or replace internal fuse. 3. Shut off timer.
VCR turns on but doesn't work.	Power LED is on but none of the controls work.	1. No cassette inserted into VCR. 2. VCR in ON timer mode. 3. VCR is paused. 4. Moisture inside VCR. 5. The VCR controls are jammed or locked out.	1. Put a cassette into the VCR. 2. Shut the timer mode off. 3. Unpause VCR. 4. Want until machine is dry. 5. Service or turn child locks off.
Remote control malfunctioning	Buttons on remote will not operate VCR functions.	1. Out of range or not pointing at IR receiver. 2. Remote batteries are dead. 3. Remote is defective or VCR circuits in need of repair. 4. Universal remote programmed incorrectly.	1. Move closer to VCR or point directly at IR receiver. 2. Replace batteries. 3. Replace remote or take VCR in for service. 4. Program remote according to manual.
Picture is unstable on playback.	The video jumps and rolls, and won't stabilize on the screen.	1. Tracking is not adjusted right. 2. Macrovision is too strong. 3. Dirty tape guides. 4. Dirty control track head. 5. Tape is defective. 6. Tape tension is incorrect.	1. Adjust the tracking. 2. Use a sync stabilizer (macro buster). 3. Clean. 4. Clean. 5. Check by inserting a tape you know is okay. 6. Service after checking with a good tape.
Picture is flagging.	The picture's top bends one way or the other.	1. Macrovision is too strong. 2. Tracking is off. 3. Tape is stretched. 4. Tape tension incorrect. 5. Controls are malfunctioning.	1. Use a sync stabilizer (macro buster). 2. Adjust the tracking. 3. Test VCR with a good tape. 4. Service after checking with a good tape. 5. Service.
Spots of the picture are being dropped.	Spots flash all over the picture.	1. Macrovision too strong. 2. Video heads are dirty. 3. Dirty tracking control head. 4. Bad cassette. 5. Drop-out compensator inoperative.	1. Use a sync separator. 2. Clean. 3. Clean. 4. Check against a good cassette. 5. Service.

Chapter 11: Troubleshooting Video Equipment

Problem	Symptom	Possible Cause	Solution
Noisy picture.	Screen has a "salt & pepper" look to it. Similar to a weak TV signal.	1. VCR heads are clogged. 2. Tracking is incorrect. 3. Bad cable/connector connections. 4. Video head defective.	1. Clean. 2. Track VCR. 3. Check all cables and connections. 4. Service.
Picture quality low.	Picture looks grainy.	1. Dirty video heads. 2. Old cassette. 3. Heads are defective.	1. Clean. 2. Check against a good cassette. 3. Service.
VCR will not play tape.	Picture is unstable, then the tape stops after a short time.	1. Tracking is off. 2. Cassette is defective. 3. Too much moisture in VCR. 4. Tape is misaligned inside the VCR. 5. Mechanical components defective. 6. Auto-stop detector defective.	1. Adjust the tracking. 2. Check against a good cassette. 3. Let VCR dry out for an hour. 4. Hit EJECT and push the tape back in. 5. Service. 6. Service.
No color in picture or color is poor.	No color, poor color or color flashes on/off.	1. B&W movie. 2. Low-quality recording. 3. Tracking is off. 4. TV color controls are off. 5. Color circuits malfunctioning in TV or VCR.	1. Check screen with a color tape. 2. Check with high-quality tape. 3. Adjust tracking. 4. Adjust color controls on TV. 5. Service.
Noisy sound on playback.	Humming or buzzing from the speakers, or missing a stereo channel.	1. Low-quality recording. 2. Cables or connectors defective. 3. Audio IN/OUTputs mixed up. 4. Audio heads are dirty.	1. Check with high-quality tape. 2. Inspect cables/connectors and replace if necessary. 3. Hook them up properly. 4. Clean.
Soundtrack unstable.	Audio speeds up, slows down or "flutters."	1. Tracking incorrect. 2. Control track head is dirty. 3. Old or stretched cassette. 4. Moisture in VCR. 5. Belts need replacing.	1. Track. 2. Clean. 3. Check with good tape. 4. Dry out VCR for an hour. 5. Service.
Distorted sound on playback.	Sound is garbled.	1. IN/OUTs mixed up. 2. Cables or connectors defective. 3. Audio heads are dirty.	1. Hook up the cables correctly. 2. Check and replace if necessary. 3. Clean.
No sound, or incorrect sound.	No sound, or you hear another soundtrack.	1. "TV" is selected instead of "VCR." 2. Audio cables disconnected, miswired or faulty.	1. Hit "VCR" as signal source. 2. Check wiring, replace if necessary.

Camcorder

PROBLEM	SYMPTOMS	POSSIBLE CAUSES	POSSIBLE CURES
Camcorder will not turn on.	Power LED off; unit won't work.	1. No power. 2. Tripped breaker or blown fuse. 3. Batteries weak or faulty.	1. Plug in camcorder or switch on wall unit. 2. Reset breaker or replace internal fuse. 3. Recharge or replace them.
Camcorder turns on but is not operational.	Power LED on; controls inoperative.	1. Camcorder is in STANDBY mode. 2. Batteries are weak.	1. Hit the OPERATION mode. 2. Recharge or replace them.
Picture is unstable.	Camcorder functions, but picture is unstable.	1. Weak batteries. 2. Camcorder defective.	1. Recharge or replace them. 2. Service.
Blotches in picture.	Spots appear on the picture.	1. Dirty lens. 2. Camcorder was exposed to bright light.	1. Clean lens. 2. Point the camcorder at a pure white card or wall for 15 minutes to correct the exposure.
No picture.	Camcorder is functioning, yet there is no picture.	1. Lens cap is on. 2. Camcorder in STANDBY mode. 3. Fade control on. 4. Electronics view defective.	1. Remove cap. 2. Hit OPERATION mode. 3. Shut the fade control off. 4. Service.
Colors are incorrect.	Camcorder is not recording the correct colors.	1. Color balance is off. 2. Auto balance color control is off.	1. Adjust color balance. 2. Turn auto balance on.
No sound.	Camcorder is not picking up sound.	1. Microphone is unplugged. 2. Sound is turned off. 3. Microphone defective.	1. Plug in microphone. 2. Turn sound switch on. 3. Service.

Video Disk Player

PROBLEM	SYMPTOMS	POSSIBLE CAUSES	POSSIBLE CURES
Player will not turn on.	Power LED off; unit won't work.	1. No power. 2. Tripped breaker or blown fuse.	1. Plug in player or switch on the wall switch.
Player turns on but is not operational.	Power LED is on but the controls are ineffective.	1. No disc inserted in player. 2. Player is set at PAUSE. 3. Player's controls are jammed.	1. Insert disc. 2. Shut off PAUSE. 3. Service.
Remote control malfunctioning.	Buttons on remote will not operate player functions.	1. Out of range or not pointing at IR receiver. 2. Remote batteries are dead. 3. Remote is defective or player circuits are in need of repair. 4. Universal remote programmed incorrectly.	1. Move closer to player or point remote directly at IR receiver. 2. Replace batteries. 3. Replace remote or take player in for service. 4. Program according to instructions.
Picture jumps around.	Picture occasionally jumps or skips.	1. Disc is dirty. 2. Disc has been scratched. 3. The player's laser lens is dirty. 4. Disc is warped.	1. Clean the disc. 2. Inspect and replace disc if necessary. 3. Clean lens. 4. Inspect and replace disc if necessary.
Poor quality video and audio.	Picture and sound quality are poor.	1. Disc is dirty. 2. Disc is scratched. 3. Lens is dirty. 4. Cables or connector hookups are wrong or defective.	1. Clean the disc. 2. Inspect and replace disc if necessary. 3. Clean lens. 4. Inspect and replace if necessary, or hook up properly.
Movie won't play.	Movie won't run when you hit PLAY.	1. Disc has come to the end.	1. Change or flip disc.

APPENDIX A

WEB ADDRESSES

Avinfo, Inc.
Audio/Video Search Manufacturers' Search Engine
http://www.avinfo.com/scripts/mfgrsrch.stm

Adaptive Circuitry
Video Glossary
http://www.telect.com/sources/glossary.htm

Aiwa America, Inc.
800 Corporate Dr.
Mahwah, NJ 07430
1-800-BUY-AIWA
http://www.aiwa.com/

Bose Corporation
Mountain Rd.
Framingham, MA 01701
1-800-WWW-BOSE
http://www.bose.com/

Canon USA, Inc.
One Canon Plaza
Bldg. C
Lake Success, NY 11042
1-800-OK-CANON
http://www.usa.canon.com/

Dolby Laboratories
100 Potrero Avenue,
San Francisco, CA 94103
1-415-558-0344
http://www.dolby.com/

Elcom Technologies
1-800-ELCOM-123
http://www.elcomtech.com/

Federal Communications Commission (FCC)
1919 M Street N.W.
Washington DC 20554
(202) 418-0200
http://www.fcc.gov/

GE
Thomson Consumer Electronics, Inc.
10330 N. Meridian St.
Indianapolis, IN 46290
1-800-336-1900
http://www.ge.com/

Goldstar Electronics
000 Sylvan Ave.
Englewood Cliffs, NJ 07632
201-816-2000
http://www.goldstar.co.kr/

Hitachi
3890 Steve Reynolds Blvd.
Norcross, GA 30093
800-448-2244
http://www.hitachi.com

Howard W. Sams & Company & Prompt Publications
2647 Waterfront Parkway, East Drive
Indianapolis, IN 46214-5781
1-800-428-7267
http://www.hwsams.com/

JVC America
41 Slater Drive
Elmwood Park, NJ 07407
800-252-5722
http://www.jvc.com/
http://www.jvc-america.com/

Kenwood
Kenwood USA Corp.
P.O. Box 22745
Long Beach, CA 90801-5745
800-536-9663

Mitsubishi Electric Sales of America, Inc.
P.O. Box 6007
5665 Plaza Drive
Cypress, CA 90630
800-344-6352
http://www.mitsubishi.com/

NEC USA, Inc.
800-338-9549
http://www.nec.com/

Nintendo of America, Inc.
http://www.nintendo.com/

Panasonic
Matsushita Electric Corporation of America
Parts & Service Locator:
(800) 545-2672
http://www.panasonic.com/

Philips Magnavox
Philips Consumer Electronic Company
1-800-531-0039
http://www.philipsmagnavox.com/

Pioneer Electronics, Inc.
Attn: Customer Service
Box 1760
Long Beach, CA 90801
800-746-6337
http://www.pioneer.com/

RCA
Thomson Consumer Electronics, Inc.
Attn: Customer Relations
600 N. Sherman Dr.
P.O. Box 6127
Indianapolis, IN 46201
800-336-1900
http://www.rca.com/

Samsung
Samsung Electronics America, Inc.
One Samsung Place
Edgewood, NJ 08824
800-767-4675, ext. 505

Sanyo Fisher Service
21314 Lassen St.
Chatsworth, CA 91311
800-421-5013
http://www.sanyoservice.com/

Sega of America
P.O. Box 8097
Redwood City, CA 94063
1-800-USA-SEGA
http://www.sega.com/

Sharp Electronics Corporation
Sharp Plaza,
Mahwah, New Jersey 07430-2135
201 529-8200
http://www.sharp-usa.com/

Sony of America
http://www.sony.com

Sony of Canada
http://www.sony

Toshiba America, Inc.
1-800-631-3811
www.toshiba.com/

WebTV Networks, Inc.
1-800-GOWEBTV
http://www.webtv.com/

Zenith Electronics Corporation
1000 Milwaukee Avenue
Glenview, Illinois 60025
(847) 391-7000
See Website for further #'s including 800 product numbers.
http://www.zenith.com/

Appendix A: Web Addresses

APPENDIX B

PRODUCT MANUFACTURERS

Manufacturer	Makes	Manufacturer	Makes
Canon	Canon	Matsushita	Kodak
			Magnavox
Goldstar	Goldstar		Nikon
			Olympus
Hitachi	Hitachi		Panasonic
	Kyocera		Philco
	Pentax		Quasar
	RCA		Sylvania
	Minolta		Technika
	Mitsubishi		
	Realistic	NEC	NEC
	Sears		
		Sanyo	Fisher
JVC	JVC		Sanyo
	Samsung		Vivitar
	Toshiba		
	Zenith	Sharp	Sharp
Matsushita	Chinon	Sony	Aiwa
	Curtis-Mathes		Fuji
			Kyocera
	Elmo		Pioneer
	GE		Ricoh
	Instant-Replay		Sony
	JC Penney		

Appendix B: Product Manufacturers

INDEX

SYMBOLS

10-EVENT PROGRAMMABLE TIMER 172
16-BIT GAME SYSTEM 133, 134
16:9 MODES 114
16X9 175
16X9 CAPABLE 172
181-CHANNEL TUNER 164, 172
2D 137
32-BIT 136
32-BIT RISC PROCESSORS 135
32-BIT TECHNOLOGY 137
32-VOICE SOUND 135
32X 135
35-MM FILM CAMERA 111
35-MM LENSES 110
3D 21, 134, 137
3D DIGITAL SOUND 135
3D EFFECT 134
3D GRAPHICS 135
3D PLUG-IN GLASSES 134
3D SOUND FIELD 46
3D VISUAL IMAGE 47
64-BIT GAME MACHINES 150
64-BIT SYSTEM 133
64DD 150
7-11 112, 131
8 MM 92, 105, 113, 164
8 MM CAMCORDER 93, 104
8 MM CASSETTE 114
8 MM TAPES 104
8-BIT CPU 133
8-BIT SYSTEM 133, 134
8-TRACK 11

A

A/B SWITCHES 179
A/V 117, 162
A/V CABLES 23, 142
A/V EUROCONNECTOR 92, 185
A/V HOME THEATER RECEIVER 163
A/V JACKS 11, 41, 92, 94, 121, 163
A/V LINES 16, 86, 143
A/V OUTPUTS 51
A/V PATCH CORD 184
A/V RECEIVER 48, 51, 160, 163, 175
A/V WIRING 55
AA BATTERIES 112
AA BATTERY CAPABILITY 175
AC-3 10, 98, 100, 101, 175
ACADEMICS 144
ACCESS CARD 79, 81
ACCESSORIES 12, 67, 117, 179
ACCLAIM 140
ACE COMBAT 2 140
ACOUSTIC REFLECTIONS 50
ACOUSTIC STIMULATION 27
ACTION GAMES 138
ACTION MOVIES 51
ACTIVE-CHANNEL SCAN 170
ACTIVITIES 131
ADAPTER 105, 134, 143
ADC 16
ADDITIVE COLOR PROCESS 29
ADDITIVE COLOR SYSTEM 29, 32
ADDRESSES 155
ADS 154
ADULT MOVIES 69
ADVANTAGES 77, 116
ADVENTURE GAMES 139
ADVERTISEMENT 151
ADVICE 61
AFTER-MARKER ITEM 60
AGC 56, 57
AGE 35
AIR 58
AIR 26, 58
ALARM 170
ALC 112
ALL MY CHILDREN 13
ALONE IN THE DARK 139
ALPHASTAR 80, 82, 83
ALTERNATE LANGUAGE SOUNDTRACK 46
ALTERNATING CURRENT 64
ALTERNATION 54
ALUMINUM 58, 76, 180
ALUMINUM DISH 172
AMATEUR 119
AMBIENT SOUND 50

AMERICA 67
AMERICAN 126
AMP 10
AMPLIFICATION 57, 64, 76
AMPLIFICATION GAIN 57
AMPLIFIER 10, 19, 45, 51, 67, 68, 159, 163, 179
AMPLITUDE 54
ANALOG 15, 21, 75
ANALOG BROADCAST SIGNALS 17
ANALOG CABLE 77
ANALOG HOME VIDEO EQUIPMENT 116
ANALOG PROBLEMS 68
ANALOG SIGNALS 80
ANALOG TV SIGNAL 86
ANALOG VIDEO 95
ANALOG-TO-DIGITAL CONVERTERS (ADC) 16
ANALYZER 127
ANCHORPERSON 51
ANIMATED CARTOON 32
ANTENNA 12, 16, 19, 23, 53, 54, 57, 58, 63, 64, 65, 67, 68, 74, 85, 143, 179, 182, 192, 199, 205
ANTENNA TYPES 65
APPLE COMPUTERS 120, 149
APPLIANCE 145
ARCADE GAMES 131, 133
ARCADE MACHINE 150
ARMS 88
ASPECT RATIO 17
ATARI 131
ATMOSPHERIC SOUNDS 48
ATTENUATORS 194
AUDIO 16, 21, 57, 58, 81, 86, 89, 92, 95, 142, 143
AUDIO BASEBAND LINES 55
AUDIO CABLES 185
AUDIO CD 11, 98, 100, 136
AUDIO CIRCUITS 114
AUDIO CONNECTORS 184
AUDIO DEMODULATORS 56
AUDIO DUBBING 172
AUDIO HEAD 88, 89, 90
AUDIO INFORMATION 53, 54, 55
AUDIO INPUTS 127
AUDIO JACK 10, 46, 143
AUDIO LINES 56
AUDIO MIX 127
AUDIO OPTIONS 112
AUDIO OUTPUTS 147
AUDIO PROGRAM (SAP) 46
AUDIO SIGNAL 51, 55, 56, 142
AUDIO SYSTEM 98
AUDIO TAPE 89
AUDIO-SIZED CD 98
AUDIO/VIDEO JACKS 121, 172
AUDITORY 27
AUTO CLOCK SETTING 173
AUTO FOCUS 109
AUTO FOCUS FEATURES 109
AUTO IRIS 176
AUTO LIGHT 176
AUTO PAUSE 176
AUTO SHUTOFF 176
AUTO TURN-ON 173
AUTO-HEAD CLEANING 173
AUTOMATIC CHANNEL SCANNER 91
AUTOMATIC GAIN CONTROL (AGC) 57
AUTOMATIC HEAD CLEANER 164
AUTOMATIC LEVEL CONTROL (ALC) 112
AVERAGE SYSTEM 8, 10, 11
AVID CINEMA 129
AVID TECHNOLOGY, INC. 120, 129
AZIMUTH 81

B

BABY BOOMERS 27
BAD PICTURE 67
BAND SPLITTERS 179, 188, 194, 196, 197
BANDWIDTH 54
BARGAIN PACKAGES 51
BASE SYSTEM 132
BASEBAND 17, 55
BASEBAND SIGNAL 55
BASEBAND VIDEO 56, 57
BASIC COMPONENTS 87
BASIC SERVICE 70
BASICS 86, 99, 133
BASS 45, 50
BASS EFFECTS CHANNEL 50
BATTERIES 117
BATTERY INDICATOR 110
BATTERY PACKS 104, 112
BATTERY-REMAINING DISPLAY 176
BEAMS 30
BEDROOM 45
BELTS 88
BEST BUY 156
BETA 11, 87
BETAMAX 87, 92, 105
BETTER BUSINESS BUREAU 155
BIG-TICKET ITEM 154
BIGSCREEN TVS 35, 38, 41, 44
BILLING 81
BINARY 95, 132
BITS 132, 133
BLACK 29
BLACK ANODIZED ALUMINUM MESH 76
BLACK COATING 44
BLACK FADER 176
BLACK LEVEL 44
BLACK LEVEL EXPANSION 167
BLACK STRETCH 167
BLACK-AND-WHITE 110
BLACK-AND-WHITE LCD SCREEN 160
BLACK-AND-WHITE PICTURES 127
BLACK-AND-WHITE TELEVISION 32
BLACK-AND-WHITE TV 29, 30, 34
BLANK TAPES 113, 129
BLOOMING 43
BLOOPERS 120
BLOWN FUSE 205

BLUE 28, 29, 32, 34, 35, 114
BLUE SCREEN 127
BLUNDERS 120
BOMBERMAN 138
BOOM 65
BOW-TIE ANTENNAS 66
BRAND NAME 59
BRANDS 154, 160, 165
BRIGHT 44
BRIGHTER PICTURE 44
BRIGHTNESS 29, 30, 44, 56, 105
BRIGHTNESS SIGNAL (Y) 30
BROADCAST 26, 47, 55, 92
BROADCAST CAMERA 106
BROADCAST QUALITY 108
BROADCAST SIGNAL 17, 55, 56, 57, 71, 85
BROADCAST STATIONS 86
BROADCAST TOWERS 54, 64, 65, 66
BROADCAST VIDEO SIGNAL 12
BROADCASTER 29, 46, 106, 124
BROADCASTING 119
BROADCASTING ANTENNA 67
BROADCASTING CHANNELS 79
BROADCASTING EQUIPMENT 116
BROADCASTING TOWER 65
BROCHURES 155, 160
BROWSER 147, 151
BUDGET 160
BUILDING BLOCKS 188
BUILT-IN DECODERS 83
BUILT-IN IRDS 83
BUILT-IN RECEIVERS 83
BUILT-IN SIGNAL AMPLIFIERS 65
BUILT-IN SPEAKERS 45, 46
BUILT-IN TUNER 167
BULLET AMPLIFIERS 193
BUSHIDO BLADE 139
BUTTONS 57, 59, 60, 123, 144, 151
BUY 154
BYTES 132

C

C 30, 56
C SIGNAL 30
C-BAND 75
C-BAND DISHES 77
C-BAND SATELLITE EQUIPMENT 76
C-BAND SATELLITES 77, 80
C-BAND TRANSPONDERS 75
C-BAND UNITS 80
C-MOUNT 110
CABINET 38
CABLE 12, 23, 26, 42, 58, 61, 63, 64, 67, 68, 70, 77, 83, 85, 86, 180, 181, 187, 199, 205
CABLE BOXES 12, 19, 53, 60, 68, 69, 91
CABLE BULLET 69
CABLE CHANNELS 93
CABLE COAXIAL 143
CABLE COMPANY 17, 55, 67, 68, 69, 77, 136
CABLE CONVERTER 63, 167
CABLE CUSTOMER 63
CABLE DECODER 198
CABLE LOOP 68
CABLE PROBLEMS 207
CABLE QUALITY 187
CABLE READINESS 11
CABLE SERVICE 159
CABLE SIGNAL 198
CABLE STATIONS 91
CABLE SUPPORT PERSONNEL 68
CABLE TV 17, 67, 68, 136
CABLE TV TUNER 91
CABLE VIDEO SIGNAL 92
CABLE-READY TV 69, 173
CABLE-TV RECEIVERS 87
CABLES 81, 123, 179, 180, 187
CADENCE 27
CAMCORDER 12, 17, 19, 55, 56, 92, 93, 103, 104, 107, 108, 109, 110, 111, 112, 113, 114, 119, 120, 121, 123, 124, 128, 164, 165, 175, 213
CAMCORDER ACCESSORIES 117
CAMCORDER FEATURES 112
CAMCORDER TECHNOLOGY 12
CAMERA 104, 114
CAMPER 41
CANADA 20, 57, 68, 82
CANDLE 108
CAPACITY 132
CAPCOM 139
CAPTION BOARD 124, 129
CARD 81, 147
CARRIER FREQUENCY 55
CARRIER WAVES 54
CARTRIDGE 132, 134, 142
CARTRIDGE GAME 142
CASABLANCA 45
CASES 117
CASING 144
CASSETTE 105, 114
CASSETTE TAPES 86, 114
CATHODE RAY TUBE 28
CATV 67
CAV 97, 98
CB SIGNAL 74
CCD 26, 104, 107, 108, 111, 114
CCD SENSOR 108
CD 95, 99, 100, 101, 114
CD AUDIO 135
CD DIGITAL QUALITY 96
CD RECORDING TECHNOLOGY 100
CD TECHNOLOGY 99, 100
CD VIDEO 98
CD-R 100
CD-QUALITY SOUND 81, 147
CD-ROM 23, 132, 135, 141
CD-ROM AUDIO 135

Index 223

CD-ROM DRIVE 136
CD-ROM UNIT 135
CD/DVD OPTICAL PICKUPS 175
CELL 95, 108
CENTER CHANNEL 48, 49, 50
CENTER SPEAKER 169
CENTRAL PROCESSING UNIT (CPU) 132
CHAMBER OF COMMERCE 155
CHANNEL AUTO-PROGRAM 170
CHANNEL BLOCK 170
CHANNEL LABELING 170
CHANNEL RECALL 71
CHANNEL REMINDER DISPLAY 171
CHANNEL SELECTION 143
CHANNEL SELECTOR DIALS 57
CHANNELS 11, 21, 41, 46, 48, 49, 50, 60,
    63, 65, 66, 67, 69,
    74, 77, 79, 81, 114, 121, 143, 164
CHARACTER GENERATOR 176
CHARACTER GENERATORS 126
CHARGED COUPLED DEVICES 26
CHARGED-COUPLED DEVICE (CCD) 107
CHILDPROOF LOCK 173
CHILDREN 136
CHILDREN'S SHOWS 70
CHROMA KEY 127, 128, 129
CHROMINANCE (C) 29, 30, 45, 56, 92
CINCH 92
CINCH CONNECTORS 16
CIRCUIT CITY 156
CIRCUITRY 58
CIRCUITS 57, 90, 107
CLARITY 116
CLARITY 42, 114
CLASSIFIEDS 82
CLEANER 144
CLOCK 11, 171
CLOCKWORK KNIGHT I & II 138
CLOSED CAPTIONING 43, 168
CLV 97, 98
CLV FORMAT 97
CLV UNIT 97
CMOS 107, 108
COAXIAL 181
COAXIAL CABLE 55, 74, 142, 143
COAXIAL CABLES 65
COAXIAL HOOKUPS 92
COAXIAL LINES 68
COAXIAL OUTPUT 101
CODE 59
CODE ENTRY REMOTE 59, 91
CODES 142
COLOR 28, 29, 30,
    32, 34, 37, 54, 56, 110, 127, 134, 135
COLOR BARS 28
COLOR BEAM 44
COLOR INFORMATION 43, 53
COLOR LCD SCREEN 160
COLOR LCDS 111
COLOR PROCESSOR 127
COLOR PURITY PROBLEMS 35
COLOR REPRODUCTION 117
COLOR SEPARATION 105
COLOR TELEVISION 20, 29, 30, 32, 33
COLOR TELEVISION SIGNALS 57
COLUMBIA 100
COMB FILTER CIRCUITRY 44, 167
COMB FILTERS 9
COMBINERS 194
COMBO UNITS 41
COMMERCIAL 151
COMMERCIAL EDITOR 173
COMMERCIAL SKIPPER 164
COMMERCIAL TELEVISION 26
COMMERCIAL-SKIP TIMER 171
COMMON PROBLEMS 207
COMMUNICATIONS 144
COMMUNITY ANTENNA TV (CATV) 67
COMPACT 100
COMPACT CAMCORDERS 107
COMPACT DISK 99
COMPACT OPTICAL DIGITAL DISK
    TECHNOLOGY 95
COMPACT VHS 105
COMPACT VHS TAPE 104
COMPANIES 34, 77, 114, 120, 123, 129,
    137, 144, 145, 160
COMPARISONS 167
COMPILATION 124
COMPILE 123
COMPLEMENTARY METAL OXIDE
    SEMICONDUCTOR (CMOS) 108
COMPONENT PLACEMENT 190
COMPONENT STEREO 45
COMPONENT VIDEO 101
COMPONENTS 33, 51, 55, 103, 144, 159,
    160, 190, 191
COMPOSITE VIDEO 56, 92
COMPOSITE VIDEO JACK 143
COMPOSITE VIDEO OUTPUTS 101
COMPOSITE VIDEO SIGNAL 17, 55, 56
COMPRESS 80
COMPRESSION 114
COMPRESSION TECHNOLOGY 47
COMPUTER 37, 83, 96, 114, 118,
    119, 120, 123, 124, 132, 141,
    142, 144, 145, 150
COMPUTER CHIP 132
COMPUTER DISK 142
COMPUTER GAME 151
COMPUTER HARD DRIVE 130
COMPUTER LANGUAGE 145
COMPUTER MONITOR 145
COMPUTER MONITORS 32, 146
COMPUTER OWNERS 138
COMPUTER POLYGON GRAPHICS 135
COMPUTER STORAGE 124
COMPUTER USERS 145, 151
COMPUTER-CONTROLLED VIDEO
    PROCESSORS 121
COMPUTING OPTIONS 12
CONCENTRIC RINGS 74
CONDUCTOR 180
CONNECTION 58, 81, 142, 144, 190, 191

CONNECTION PROBLEMS 207
CONNECTORS 16, 92, 121, 144, 179,
    180, 181, 187
CONSOLE 132, 134, 138, 142, 143
CONSOLE ADVANCES 137
CONSOLE EMULATION 141
CONSOLE GAMES 141
CONSOLE GAMING 150
CONSOLE UNIT 132
CONSOLES 144
CONSTANT ANGULAR VELOCITY (CAV) 97
CONSTANT LINEAR VELOCITY (CLV) 97
CONSUMER 27, 124, 153
CONSUMER GROUP 69
CONSUMER REPORTS 154
CONSUMING 153
CONTACTS 144
CONTENT 70
CONTRAST 44
CONTRAST LEVELS 44
CONTROL 27, 44, 58
CONTROL FEATURES 61
CONTROL FUNCTIONS 60
CONTROL SWITCH 205
CONTROL-L 124, 128
CONTROL-M 124
CONTROL-S 124
CONTROL-T 124
CONTROLLER 123
CONVERTER FEATURES 70
COOKING 41
COPIERS 141, 142
COPIES 208
COPPER 58, 180
COPPER LINES 179
COPY 201
COPY-PROTECT 208
COPYRIGHT LAWS 100, 141
CORNER YAGI ANTENNA 66
COST 51, 70, 79, 125, 129, 149
CPU 132, 133
CREDITS 126
CROSS-SECTION CIRCULAR DISH 74
CRT 28
CURRENT 54
CURVED SCREEN 45
CUSTOMER 71
CUTS 127, 128
CUTTING 129
CYCLE 54

D

DAC 16
DAEWOO 79
DARK 44
DARK SCREENS 167
DAT 114
DATA 54, 132
DATA COMPRESSION 126
DATA DOWNLINKS 83
DATABASE TOOLS 150

DATE/TIME DISPLAY 176
DAYLIGHT 35
DBS 12, 17, 42, 61, 63, 70, 73, 74,
    77, 80, 82, 83, 160, 164, 172
DBS COMPANIES 79
DBS DISHES 81
DBS EQUIPMENT 80
DBS SATELLITE 76, 80
DBS SATELLITE DISHES 77
DBS SERVICES 61, 78, 79, 81, 83, 164
DBS SIGNALS 93
DBS SYSTEM 68, 80, 83, 159
DBS UNITS 81
DECIBELS 64, 192
DECODER 10, 19, 47, 48, 49, 51, 52, 83
DECODER BOXES 68, 69
DECODER SIGNAL SELECTOR 198
DECOR 38
DEDICATED REMOTE 59
DEEP FRINGE 65
DEEPEST FRINGE 65
DEFINITIONS 132
DEGAUSSING 35
DEGAUSSING CIRCUITS 35
DEMODULATION 54, 55
DEPARTMENT STORE 156
DESCRAMBLER BOXES 68
DESCRAMBLERS 68, 77
DESIGN 65
DESKTOP COMPUTER 96
DESPOSITO, JOE 94
DETACHABLE LENSES 110
DETAIL 27, 41, 44
DIAGONAL MEASUREMENT 17, 38
DIALOGUE CHANNEL 50
DIALS 57
DIGITAL 15,
    21, 71, 75, 80, 86, 92, 98, 99, 104,
    113, 117, 160
DIGITAL (AC-3) DECODER 169
DIGITAL AUDIO TRACKING 164
DIGITAL AUTO TRACKING 173
DIGITAL CAMCORDERS 113, 114, 116
DIGITAL CAMERAS 115, 120
DIGITAL COMPRESSION TECHNOLOGY
    99
DIGITAL CONTROL 171
DIGITAL DATA 99
DIGITAL DATA DEPOT 87
DIGITAL EDITING 120, 124
DIGITAL EDITING PACKAGES 124
DIGITAL EDITING VCR 128
DIGITAL FRAME STORAGE 98
DIGITAL INFORMATION 58, 95
DIGITAL MANIPULATION 114
DIGITAL MICROPROCESSOR 16
DIGITAL PROCESSING SYSTEMS, INC.
    129
DIGITAL PROCESSOR CHIP 135
DIGITAL PRODUCT 77
DIGITAL SATELLITE RECEIVERS 12
DIGITAL SATELLITE SYSTEM (DSS) 17, 77

DIGITAL SATELLITE TRANSPONDERS 80
DIGITAL SIGNAL 70, 80, 81
DIGITAL SIGNAL PROCESSING (DSP) 169
DIGITAL SIGNAL PROCESSORS 150
DIGITAL SOUND 113, 135
DIGITAL STORAGE 114
DIGITAL SURROUND SOUND 10, 51
DIGITAL SYSTEMS 120
DIGITAL TECHNOLOGY 100, 120
DIGITAL TELEVISION (DTV) 9, 18, 39, 58
DIGITAL TV-TOP DECODER BOX 79
DIGITAL UNITS 12
DIGITAL VIDEO 101, 113
DIGITAL VIDEO DISK 19, 42
DIGITAL VIDEO DISK PLAYERS (DVD) 11, 98
DIGITAL VIDEO EQUIPMENT 160
DIGITAL VIDEO NOISE REDUCTION CONTROL 175
DIGITAL VIDEO PICTURES 100
DIGITAL VIDEO PROCESSOR 127, 135
DIGITAL VIDEO SYSTEMS 121
DIGITAL WHITE BALANCE 176
DIGITAL ZOOM 109
DIGITAL-TO-ANALOG CONVERTER (DAC) 16, 175
DIGITALLY-RECORDED MOVIES 95
DIGITIZE 119
DIMENSIONAL EFFECTS 49
DIMENSIONAL SOUND EFFECT 48
DIRECT BROADCAST SATELLITE (DBS) 12, 17, 77, 172, 184, 201
DIRECT BROADCAST SATELLITE TECHNOLOGY 77
DIRECT WAVE SIGNAL 64
DIRECT-TO-HOME SERVICE (DTH) 79
DIRECTION 57
DIRECTIONAL CAPABILITY 49
DIRECTV 79, 81, 82, 164
DISADVANTAGES 77
DISCO 11
DISH 58, 74, 76, 77, 79, 81, 83
DISH ANTENNA 81
DISH ATTACHMENT ANTENNAS 67
DISH NETWORK 80, 82
DISH NETWORK 79, 164
DISK 95, 96, 98, 99
DISK DRIVES 142
DISK SPACE 129
DISSOLVES 126, 127
DISTORTION 45
DISTRIBUTION AMPLIFIERS 194
DISTRICT REPRESENTATIVE 160
DOLBY 18, 47, 48, 50, 98, 100, 101, 135
DOLBY 3 163
DOLBY 3 STEREO MODE 49, 169
DOLBY DIGITAL 10, 11, 19, 48, 50, 51, 96, 98, 100, 101, 175
DOLBY DIGITAL SURROUND 10
DOLBY DIGITAL (AC-3) UNIT 51
DOLBY LABORATORIES, INC 18, 47
DOLBY PRO LOGIC 10, 19, 50, 51, 100

DOLBY PRO LOGIC SURROUND 50
DOLBY PRO LOGIC A/V RECEIVER 51
DOLBY PRO LOGIC DECODERS 49
DOLBY, RAY 18
DOLBY SURROUND 19, 45, 48, 49, 50, 51, 98, 169
DOLBY SURROUND AC-3 50
DOLBY SURROUND PRO LOGIC 48
DOLBY SURROUND PRO LOGIC DECODER 48
DOLBY SURROUND SIGNAL 48
DOT 34
DOUBLE IMAGES 64
DOUBLE-LAYERED DISKS 101
DOUBLE-SIDED DISKS 101
DOUBLE-SIDED PLAY 98
DOWNLINK 75
DOWNLINK SIGNAL 75
DPS 129
DSS 17, 74, 77, 79, 81, 82, 83, 164
DSS COMPATIBLE DISH 172
DSS PROGRAMMING 164
DSS SATELLITE DISH 78
DSS SYSTEM 81
DTV 17, 18, 27, 39, 43, 44, 83
DTV BROADCAST STANDARD 50
DTV FORMAT 32
DTV STANDARDS 27, 83
DUAL QUICK SET 174
DUAL-MODE PCM 114
DUAL-OUTPUT LNB 172
DUNGEONS AND DRAGONS 140
DUPLICATE FUNCTIONS 161
DVD 11, 12, 19, 95, 97, 98, 99, 100, 101, 159, 160, 175
DVD MOVIES 101
DVD PLAYER 11, 98, 100
DVD STANDARD 50
DVD TECHNOLOGY 43

E

E.A. SPORTS 140
EARS 46
EARTH 71, 72
ECHOSTAR 80, 82, 83
ECONOMY PRODUCTS 157
EDIT JACK 175
EDIT-CONTROL FEATURE 124
EDITING 110, 120, 129
EDITING CONTROLLER 128
EDITING CONTROLLERS 123
EDITING EQUIPMENT 124
EDITING METHODS 121
EDITING PRINCIPLES 119
EDITING SOFTWARE 130
EDITING SYSTEM REQUIREMENTS 129
EDITING VCRS 93, 128
EDUCATIONAL DISKS 96
EFFECTS 48, 119, 129
EIDOS INTERACTIVE 139
EIGHT MILLIMETER 92

EJECT 173
ELECTRICAL FIELDS 44
ELECTRICITY 180
ELECTRODES 33
ELECTROMAGNET 86, 88
ELECTROMAGNETIC ENERGY 54
ELECTROMAGNETIC WAVES (EM) 21, 23, 54
ELECTROMAGNETS 33, 89
ELECTROMECHANICAL IMPULSES 103
ELECTRON BEAM 33, 34, 56
ELECTRON GUN 30, 31, 32
ELECTRON GUN ASSEMBLY 33
ELECTRONIC 44
ELECTRONIC ARTS 140, 141
ELECTRONIC EQUIPMENT 205
ELECTRONIC GEAR 59
ELECTRONIC IMAGE STABILIZER 176
ELECTRONIC IMAGING DEVICE 107
ELECTRONIC INFORMATION 144
ELECTRONIC SIGNAL 12, 26, 52
ELECTRONIC VIEWFINDERS 110
ELECTRONICS 77, 90
ELECTRONICS BOUTIQUE 156, 161
ELECTRONICS FEATURES 154
ELECTRONICS JARGON 157
ELECTRONICS NOW 154
ELECTRONICS STORES 10, 156, 163
ELECTRONS 30, 33
ELECTROSTATIC FIELDS 33, 44
ELEVATION 81
EM SPECTRUM 54
EM WAVE 58
EM WAVES 54
EM WAVES 54, 58, 66
EMAIL 19, 144, 145, 148
EMISSIVE DISPLAYS 37
EMULATOR 141
ENCODE 48
ENCODED SIGNAL 48
ENCODING/DECODING SYSTEM 50
END OF TAPE 176
ENGINEERS 57, 99
ENGLAND 126
ENGLISH LANGUAGE TRACK 47
ENHANCED STEREO 50, 51, 169
ENHANCED STEREO UNITS 52
ENHANCEMENT 50
ENTERTAINMENT CABINET 38
ENTERTAINMENT CENTERS 14
ENTERTAINMENT PACKAGES 150, 151
ENTERTAINMENT SYSTEM WIRING 180
EP (EXTENDED PLAY) 105
EQUATOR 72
EQUIPMENT 51, 56, 59, 79, 124, 129, 161, 165, 190
EQUIPMENT DUPLICATION 87
EQUIPMENT INPUTS 190
EQUIPMENT OUTDATING 138
EQUIPMENT OUTPUTS 190
EQUIPMENT RENTALS 70
ERASE 90

ERASE HEAD 88, 89
ERRORS 114
ESTIMATE 207
EUROCONNECTOR 185
EUROPEAN 20
EUROSAT 79
EVANS, ALVIS J. 67
EVEN LINES 43
EVENTS 63, 69
EXPANDED HOME ENTERTAINMENT CENTER 7
EXPANDED SERVICE 70
EXPANDED SYSTEM 8, 10, 11
EXPANSION TECHNOLOGY 48
EXTERIOR COLOR 38
EXTERNAL BUTTON 91
EXTERNAL MICROPHONE 112
EXTERNAL MICROPHONE JACK 176
EXTERNAL STEREO 47

F

F-22 INTERCEPTOR 140
F-CONNECTORS 181, 185
FACEPLATE 34
FACTORY WARRANTY 206
FADES 119, 128, 129
FAMILY ROOM 40
FAST MOTION 173
FAULTY ADAPTERS 143
FAXES 144
FCC 39, 71, 81
FCC SIGNAL STANDARDS 61
FEATURES 41, 111
FEATURES 9, 40, 41, 45, 52, 87, 94, 98, 100, 111, 113, 154, 164, 167
FEEDHORN 74, 76
FEMALE 180
FEMALE COAXIAL CONNECTOR 142
FIBER OPTIC DIGITAL NETWORK 71
FIBER OPTIC LINES 58
FIBERGLASS 76
FIBERGLASS COATED SOLID DISH 76
FIELDS 32
FIGHTING GAMES 139, 151
FILES 145
FILM INDUSTRY 119
FILMING TO EDIT 120
FILTERING 57
FILTERS 110, 194
FINAL FANTASY 140
FINE-TUNING CONTROLS 65
FIRST SCENE 113
FISHER 79
FIXED FOCUS 109
FIXED ORBIT 72
FLAT SCREENS 45, 168
FLAT-FRONT SCREEN 35
FLAT-LEAD 182
FLAT-LEAD CABLE 182
FLAT-SCREEN PICTURE TUBE 45
FLICKER 32

FLINTSTONE 131
FLIP FUNCTIONS 127
FLOPPY DISK 142
FLYING ERASE HEAD 90, 93, 107, 128, 173
FM 65, 66
FOCUS 35, 44
FOCUS SETTING 109
FONTS 126
FOOTAGE 120, 121, 124, 129
FOOTAGE TAPE 121, 129
FOOTCANDLE 108
FOOTPRINT 74, 83
FOREIGN LANGUAGE AUDIO TRACK 47
FORMAT 88, 96, 97, 98, 103, 104, 113, 164
FOUR-HEAD UNIT 11
FOUR-HEAD VCR 93
FRAME 32, 43, 97, 111
FREQUENCY 54, 65, 71, 74, 75, 80
FREQUENCY BAND 54, 57, 75
FREQUENCY MODULATED (FM) 71
FREQUENCY SYNTHESIZED PLL TUNERS 57
FRIENDS 154
FRINGE 65
FRISBEE 77
FRONT JACKS 94
FRONT JACKS 161
FRONT PROJECTION TVS 37
FRONT SPEAKERS 169
FRONT STEREO SPEAKERS 50
FRONT-FIRING SPEAKERS 169
FULL TV-TOP KEYPAD DISPLAY 70
FULL-MOTION PLAYBACK 135
FULLY DIGITAL 100
FUTURE 150

G

GADGETS 203
GAIN 57, 64, 76
GAME 131, 132, 133, 134, 135, 136, 138, 142, 150
GAME CARD 134
GAME CARTRIDGE 23, 132, 141
GAME CARTRIDGE PORT 148
GAME COMPANIES 137
GAME COMPANY 151
GAME CONSOLE 41, 142, 143, 150, 151
GAME ENVIRONMENTS 137
GAME GEAR 134
GAME SAVING 142
GAME SOFTWARE 132
GAME SYSTEM 143
GAMEBOY 133, 141
GAMING TECHNOLOGY 136
GAMING WORLD 131
GAMMA RAYS 54
GARABEDIAN, KEVIN 94
GAS-PLASMA 37
GAS-PLASMA SCREENS 37
GE 9, 79, 157
GEARS 88
GENERAL AUDIENCE 70
GENESIS 135, 136, 138, 140
GENESIS 32X 135
GEOSTATIONARY ORBIT 72
GEOSYNCHRONOUS ORBIT 72
GHOST ELIMINATOR 64, 194
GHOSTING 64
GIGAHERTZ 54
GLARE 35, 45
GLASS 44
GLASS FACEPLATE 34
GLOSSARY 7
GO-TO FEATURE 174
GOLD DISK, INC. 123
GOVERNMENT ACCESS 70
GRAINY LOOK 42
GRAPHICS 133
GREEK 26
GREEN 28, 29, 32, 34, 35, 114
GROUND 81

H

HARD COPIES 145
HARD DRIVE 119, 124
HARD DRIVE SPACE 125, 129
HARDWARE 67, 144, 132, 179
HARDWARE MANUFACTURERS 79
HARDWARE STANDARDS 50
HBO 68
HD 124
HDTV 17, 18, 27, 39, 43, 44, 71, 162
HDTV FORMATS 43
HDTV SIGNALS 35
HEADS 88, 90
HEIGHT 54, 67
HELICAL SCAN 90
HERTZ 54
HI-FI 19, 89, 161, 163
HI-FI SIGNAL 89
HI-FI STEREO 11, 93, 113, 174
HI-FI STEREO AUDIO HEADS 89
HI-FI STEREO SOUND 164
HI-FI VCR 11
HI8 12, 42, 96, 106, 108, 113, 116, 118, 119, 164
HI8 FORMATS 114
HIGH DEFINITION TELEVISION (HDTV) 18, 39
HIGH FIDELITY 19
HIGH FREQUENCY CARRIER WAVE 54
HIGH-BAND CAMCORDERS 105
HIGH-DEFINITION MOVIE 98
HIGH-FIDELITY VCR 93
HIGH-ISOLATION SWITCHES 189
HIGH-RESOLUTION CAMCORDER 164
HIGH-RESOLUTION EQUIPMENT 159
HIGH-RESOLUTION IMAGES 11

HISTORY 86
HITACHI 79
HOLES 95
HOLIDAY SHOPPING SEASON 157
HOLLYWOOD 119
HOME 5, 65, 81
HOME COMPUTERS 128
HOME ENTERTAINMENT CENTER 7, 19, 40, 63
HOME ENTERTAINMENT PACKAGE 63
HOME ENTERTAINMENT SYSTEM 23, 145, 151, 179
HOME GAME MARKET 150
HOME MOVIE 121
HOME MOVIE CAMERA 119
HOME MOVIES 12, 108, 124, 130
HOME SHOPPING NETWORKS 70
HOME STEREO 10
HOME THEATER 10, 45, 46, 48, 101
HOME THEATER A/V RECEIVERS 169
HOME THEATER EQUIPMENT 50
HOME THEATER MARKET 65
HOME THEATER ROOM 163
HOME THEATER SOUND PACKAGE 10
HOME THEATER SOUND RECEIVER 19
HOME THEATER SOUND SYSTEM 51
HOME THEATER STEREO 47
HOME VIDEO 120
HOOKUP 203
HOOKUP DIAGRAMS 196
HOOKUP METHOD 142
HOOKUP PROCEDURES 67
HORIZON 76
HORIZONTAL 32, 56
HORIZONTAL AXIS 42
HORIZONTAL BLANKING PULSES 56
HORIZONTAL RESOLUTION 42, 43, 92
HOWARD W. SAMS & COMPANY COMPLETE VCR TROUBLESHOOT 94
HOWARD W. SAMS COMPLETE GUIDE TO AUDIO 52
HTML 3.0 147
HUDSON 138
HUE 30
HUGHES ELECTRONICS 80
HUGHES ELECTRONICS CORP 77, 81
HUGHES NETWORK SYSTEMS 80
HUMAN EYE 32
HYBRID SATELLITES 75

I

IBM 149
IF 56, 57
ILLUMINATION 40
IMAGE STABILIZATION 111
IMAGE STABILIZATION CIRCUITS 112
IMAGES 30, 32, 44, 111
IMAGING DEVICES 104, 107, 108
IMAGING PROCESS 107
IMAGING RESOLUTION 108
IMPEDANCE 64, 187

IMPEDANCE-MATCHING TRANSFORMER 65
IMPROVEMENTS 27, 159
IN-CAMCORDER EDITING 120
IN-HOME VCR MECHANICAL REPAIR & CLEANING GUIDE 94
IN-LINE AMPLIFIERS 193
INCOMING SIGNAL 74
INCREDIBLE UNIVERSE 156
INDEX POINTS LOCATOR 174
INDEX SEARCH 174
INDICATORS 174
INDO-EUROPEAN 26
INDOOR ANTENNAS 65
INFOMERCIALS 6, 67
INFORMATION 43, 53, 55, 58, 63, 83, 114
INFORMATION PEDDLERS 68
INFRARED BEAM 109
INFRARED KEYBOARD 144
INFRARED RECEIVER 59
INFRARED REMOTE 60
INFRARED TRANSMITTER 59
INPUT/OUTPUT LINE 92
INPUTS 127, 190
INSTALLATION 78, 81, 165
INSTALLATION FEES 70
INSTALLING COMPONENTS 191
INSTANT REPLAY 168
INSULATION 184
INTEGRATED CIRCUIT 107
INTEGRATED ELECTRONICS 94
INTEGRATED MICROPROCESSORS 150
INTEGRATED RECEIVER/DESCRAMBLER (IRD) 76
INTEGRATED REMOTE 59
INTEL 150
INTENSITY 108
INTERACTIVE DISK GAMES 96
INTERFACE 27, 58, 61
INTERFERENCE 80, 184, 205, 208
INTERLACING 32
INTERMEDIATE FREQUENCY (IF) 57
INTERNAL BAND SPLITTER 198
INTERNAL SCREEN 35
INTERNET 12, 19, 20, 141, 144, 145, 147, 148, 149, 155
INTERNET ACCESS 71, 83
INTERNET APPLIANCES 12, 19, 144, 147, 150
INTERNET CONNECTION 150
INTERNET CONSOLES 144, 145
INTERNET DEVICES 145, 150
INTERNET DOWNLINKS 83
INTERNET PROVIDERS 148
INTERNET SERVICES 83
INVAR MASK 44
INVAR SHADOW MASKS 35
IR WIRELESS REMOTE 70
IRC (INTERNET RELAY CHAT) 148
IRD 76, 78, 80, 81, 83
ISP (INTERNET SERVICE PROVIDER) 148

Index 229

J

JACKS 16, 112, 162, 163, 168
JAPAN 20, 57
JAPANESE 126
JARGON 154, 157
JAVA 145
JAVA APPLICATIONS 149, 150
JAVA PROGRAMS 145
JAVA WORD PROCESSOR 145
JEOPARDY 35
JLIP 176
JOG 174
JOG/SHUTTLE 11, 93
JOHN MADDEN FOOTBALL 140
JOINT LEVEL INTERFACE PROTOCOL (JLIP) 176
JOYSTICKS 131, 132, 143, 144
JVC 87, 104, 105, 113, 116, 117

K

K-BAND 75
K-MART 117, 128, 141, 157
KBPS (KILOBITS PER SECOND) 147
KELVIN 76
KEYBOARD 149
KIDS' ROOM 41
KILO 132
KILOBYTES 132
KINGS QUEST 139
KITCHEN 41, 45
KNOBS 57
KNOWLEDGE 154
KU-BAND 75, 80
KU-BAND SATELLITE SYSTEMS 77
KU-BAND SATELLITE TECHNOLOGY 77
KU-BAND SIGNAL 80

L

LANC 124
LANDS 95
LANDSCAPES 137
LAPTOP 96
LAPTOP COMPUTER SCREENS 110
LARGE SCREEN TELEVISION 11
LASER 95, 98, 99
LASER BEAM 99
LASER DISK 11, 42
LASER DISK PLAYER 11, 19, 202
LASER TECHNOLOGY 100
LASER VIDEO DISK 95, 96, 97
LASER VIDEO DISK PLAYERS 95
LASERDISK 95, 98
LASERDISK PLAYER 95, 98
LASERDISK TECHNOLOGY 99
LATE-NIGHT MOVIE 41
LATERAL RODS 65
LATIN 6
LCD 37, 134, 164
LCD DISPLAY 60, 104, 133
LCD SCREEN 110, 164
LCD VIEWFINDER 164
LCD WALL UNIT 35
LCD-VIEWFINDER CAMCORDER 111
LEARNING MODE 60
LEARNING TOOLS 136
LED SIGNAL STRENGTH METER 81
LEFT STEREO CHANNEL 48
LEMMINGS 141
LENS 104, 108, 111
LETHAL WEAPON 13
LETTERBOX 37, 175
LETTERBOX 100
LICENSE 48
LICENSED TECHNOLOGY 50
LIGHT 26, 31, 32, 37, 54, 107, 108, 111, 114
LIGHT SENSORS 44, 168
LIGHT SPEED 54
LIGHT SYSTEM 7, 9, 10, 11
LIGHT VALVE 37
LIGHTED BUTTONS 60
LIGHTING 117, 134
LIGHTING SOURCE 104
LINE-LEVEL 56
LINEAR EDITING 124, 125, 129
LINEAR SYSTEM 128
LINES 32, 58, 142
LINES OF RESOLUTION 168
LIQUID CRYSTAL DISPLAY (LCD) 110
LIVING ROOM 40, 50
LNB 74, 76, 81
LNB/IRD 201
LOCAL BROADCAST 85
LOCAL CHANNELS 12, 70, 83, 159
LOG PERIODIC ANTENNAS 66
LOG PERIODIC UNIT 67
LOOSE PARTICLES 144
LOW-LIGHT SCENES 108
LOW-NOISE BLOCK DOWNCONVERTER (LNB) 74
LOW-RESOLUTION SIGNAL 44
LP 105
LP ALBUM 95
LUCAS, GEORGE 50, 118, 119
LUMINANCE 29, 34, 45, 56, 92
LUX 108

M

MAC OS 145
MACINTOSH 128, 145
MACRO 109
MACROVISION 208
MAGNAVOX 59, 80
MAGNAVOX WEBTV 148
MAGNETIC DATA 88
MAGNETIC FIELDS 44
MAGNETIC INFORMATION 90
MAGNETIC MEDIA 86
MAGNETIC READ/RECORD VIDEO HEADS 88
MALE 180

MALE CONNECTORS 184
MANUAL 87, 121
MANUAL EDITING 121
MANUAL OVERRIDE 109
MANUFACTURE 48
MANUFACTURER 44, 59, 66, 79, 83, 93, 109, 114, 124, 133, 145, 154, 155, 160, 161, 167, 206
MANUFACTURER MODELS 147
MASS-MARKET TV SET 41
MASS-MARKETERS 157
MAST 65
MAST MOUNT 193
MASTER BEDROOM 41
MASTER RECORDING TAPE 124
MATCHING TRANSFORMERS 179, 188
MATRIX DISPLAYS 37
MATSUSHITA 80
MECHANICAL 44
MECHANICAL COMPONENTS 88
MECHANICAL ZOOM 109
MEDIA 77
MEDIA TECHNOLOGIES 11
MEDIUMS 58
MEG 132
MEGA 132
MEGABYTES 132, 145, 147
MEGAHERTZ 54
MEMCORP 79
MEMOREX 79
MEMORY 12, 125, 136
MEMORY CHIP 113, 132
MEMORY SIZE 132
MENU OPTIONS 58
MENU SYSTEMS 83
METAL 76
METAL CYLINDER 88
METAL MESH 76
METAL SHIELD 184
METAL-EVAPORATED TAPE (METAL-E) 106
METAL-OXIDE SEMICONDUCTOR (MOS) 107
METALLIC MATERIAL 86
METALLIC MEDIUM 58
MEXICO 20, 57
MGM/UA 100
MICRO-SATELLITE DISH ANTENNAS 77
MICROPHONES 21, 46, 112
MICROPROCESSOR 145, 206
MICROPROCESSOR TECHNOLOGY 60
MICROSOFT 150
MICROSOFT INTERNET EXPLORER 3.0 147
MICROWAVE 41, 75
MICROWAVE BAND 54, 75
MICROWAVE FREQUENCY RANGE 74
MIDI SOUND PROCESSOR 147
MIDWAY 139
MIME-ENCODED FEATURES 147
MINI HOME ENTERTAINMENT CENTERS 13

MINI-DV 113
MINIATURE COMPUTER 149
MINIATURE SCREENS 110
MINIATURIZATION 103
MIRO COMPUTER PRODUCTS AG 129
MIRRORS 111
MITSUBISHI 100
MODEL FEATURES 167
MODELS 88, 96, 160
MODEM 145, 147
MODERN CIVILIZATION 5
MODERN TELEVISIONS 44
MODULATED SIGNAL 92, 187
MODULATION 54
MONITOR 25, 110
MONO 56, 112, 163
MONO AUDIO TRACK 89
MONO EFFECT 47
MONO RECORDING 89
MONO SOUND 46
MONO SOUND UNIT 11
MONO SPEAKERED TVS 52
MONO SPEAKERS 10
MONO VCR 11, 90, 200
MONO-SURROUND CHANNEL 50
MONOPOLY 69
MORTAL KOMBAT 14, 139
MOS 107
MOS SENSOR 108
MOSAIC COMPUTER INC. 205
MOTHERBOARD 128
MOTION SENSOR 177
MOTORIZED MOUNTINGS 76
MOTORS 76, 88
MOUNTABLE LIGHT 117
MOUNTING KIT 81
MOUNTINGS 192
MOUSE 149
MOVIE 48, 63, 69, 85, 108, 129, 151
MOVIE CHANNELS 70
MOVIE COMPANIES 100
MOVIE DIRECTOR 45, 119
MOVIE RENTAL STORES 11
MOVIE SCENES 45
MOVIE THEATER 7, 10, 50, 52
MOVIE THEATER SOUND SYSTEMS 10
MOVING PICTURES EXPERTS GROUP (MPEG) 99, 126
MPEG 99, 126
MPEG-2 DECODER 99
MPEG-2 DECOMPRESSION 175
MTS 46, 47, 51, 163
MTS DECODER 91
MTS STEREO 162
MTS STEREO RECEIVER 10
MTS/SAP 51, 163
MULTICHANNEL STEREO SOUND TECHNOLOGY 48
MULTICHANNEL TELEVISION SOUND 47
MULTICHANNEL TELEVISION SOUND DECODER (MTS) 46
MULTILINGUAL MENUS 171

MULTIPLAYER GAMING 150
MULTIPLE LANGUAGE TRACKS 101
MULTIPLE LANGUAGES 126
MULTIPLE RECORDING SOURCES 46
MULTIPLE SPEAKERS 46
MULTIPLE TUNERS 168
MULTIPLE VIDEO OUTPUTS 147
MULTIPLE-FORMAT TAPES 121
MURPHY'S LAW 205
MUSIC 70, 129

N

N-64 23
N64 134, 140
NAMCO 139, 140
NARROW-BEAM TRANSPONDER TECHNOLOGY 83
NATIONAL TELEVISION STANDARDS COMMITTEE (NTSC) 20, 57
NBA JAM 140
NC 149
NECK 33
NEGATIVE PICTURES 127
NEGATIVE/POSITIVE TRANSPOSE 177
NES 133, 134, 138, 140, 141
NET 151
NETLINK UNIT 148
NETSCAPE 147, 149
NETSCAPE 3.0 147
NETWORK 149
NETWORK COMPUTER, INC. 150
NETWORK COMPUTERS (NCS) 149
NEUTRAL 149
NEUTRAL SCREEN 35
NEWS 41, 70
NEWSPAPER 70, 82
NEWTEK, INC. 129
NINTENDO
  41, 133, 136, 137, 138, 140, 141
NINTENDO 64 (N64) 134
NINTENDO 64 DISK DRIVE (64DD) 150
NINTENDO ENTERTAINMENT SYSTEM (NES) 23, 131, 133
NOISE 75
NOISE LEVELS 116
NOISE TEMPERATURE 76
NOISE-REDUCTION CIRCUITRY 100
NON-EMISSIVE DISPLAYS 37
NON-INTERLACE SCHEMES 32
NONLINEAR EDITING 124, 125, 130
NONLINEAR SYSTEM 128
NORMAL 65
NORTH AMERICA 20, 126, 147
NTSC (NATIONAL TELEVISION STANDARDS COMMITTEE) 20, 57, 126
NTSC SIGNAL 43, 58
NTSC TV 32
NTSC TV SET 56
NTSC VIDEO SIGNAL STANDARD 27
NUMBERS 114

O

OBSOLETE CONSOLE 138
OBSTRUCTIONS 65
ODD LINES 43
ON-DISH LED SIGNAL FINDER 172
ON-SCREEN COLORS 134
ON-SCREEN CONTROL 60
ON-SCREEN DISPLAY 169
ON-SCREEN MENU PROGRAM 83
ON-SCREEN MENU SYSTEMS 83
ON-SCREEN MENU-ASSISTED PROGRAMMING 91
ON-SCREEN MENUS 171, 172
ON-SCREEN PROGRAMMING 11, 60, 61, 91, 93, 161, 162, 164, 170
ON-SCREEN SERVICES 61
ONE-BUTTON RECORD 172
ONE-DIMENSIONAL SOUND 46
OPERATING SYSTEM 129
OPERATION 98
OPTICAL DISK TECHNOLOGY 95
OPTICAL ILLUSION 30
OPTICAL OUTPUT 101
OPTICAL VIEWFINDER 110, 111
OPTIMIZING VIDEO SIGNALS 187
ORACLE CORP. 149
ORACLE NC, INC. (NCI) 149
ORIGIN 140
ORION RISC CPU 147
OS 7.5X 129
OS/2 145
OUT 121
OUTAGES 69
OUTDOOR ANTENNA 65
OUTLETS 189
OUTPUT CHANNEL 143
OUTPUT SIGNAL 52
OUTPUTS 190
OVERLAY 60

P

PACKAGE DEAL 79
PACKAGES 70
PACKAGING 79
PAINT 127
PAL (PHASE ALTERNATE LINE) 20
PALMCORDERS 17, 104, 112
PAN 100
PAN-AND-SCAN 37, 175
PANASONIC 80, 100, 113
PANNING 118
PARABOLIC CIRCULAR DISH 74
PARENTAL CONTROL 71
PARENTAL LOCKOUT CONTROLS 101
PAST 107
PASTING 129
PATCH CORDS 180, 205
PAUSE 121
PAY TV 70
PAY-PER-COMPUTER PROGRAMS 83

PAY-PER-PROGRAM-USE SCHEMES 83
PAY-PER-VIEW 69, 70, 71, 81
PC 120, 128, 136, 141, 149, 150
PCM 114
PCM STEREO 175
PENTIUM 129
PERFORMA 120
PERIPHERAL 123, 145
PERSONAL COMPUTER
    119, 141, 145, 149, 150
PERSONAL VIDEO SYSTEM 6
PHANTASY STAR 140
PHANTOM CENTER 21
PHANTOM CENTER CHANNEL 48, 49
PHANTOM DIALOGUE CHANNEL 48
PHANTOM LINE 31
PHASE-LOCKED-LOOP TUNERS (PLL) 57
PHILIPS 80
PHILIPS/MAGNAVOX 147
PHONE JACK 148
PHONE LINES 81, 147
PHONE MODEM 83
PHONE NUMBERS 155
PHONES 144
PHOSPHOR 30, 34, 35, 44
PHOSPHOR COATING 35
PHOSPHOR ELEMENT 44
PHOSPHOR SCREENS 35
PHOSPHOR-COATED BARS 34
PHOSPHOR-COATED GLASS SCREEN 30
PHOSPHOR-COATED SCREEN 32
PHOTO MONTAGES 109
PHOTOCELL 95
PHOTOELECTRIC CELL 95
PHOTOGRAPHIC FILM 107
PICTURE 30, 32, 34, 42, 44, 54, 60, 65,
    72, 80, 141
PICTURE HEIGHT 42
PICTURE INFORMATION 53, 54, 56, 90
PICTURE QUALITY 12, 27, 34
PICTURE RESOLUTION 61
PICTURE SCREEN 34
PICTURE TECHNOLOGY 35
PICTURE TUBE 28, 33, 38
PICTURE TUBE
    26, 27, 28, 34, 35, 38, 41, 52, 55, 107
PICTURE TUBES 35
PICTURE WIDTH 42
PICTURE-IN-PICTURE (PIP) 8, 168
PICTURE-IN-PICTURE SET 154
PIONEER 100
PIP 8, 9, 162
PIRATES 68
PITS 95
PIXELS 28, 37, 108
PLASTIC 88
PLASTIC TAPE 86
PLAY 122
PLAY BUTTON 87, 98
PLAY, INCORPORATED 116
PLAY-ALONG STORIES 136
PLAYBACK 21, 95
PLAYBACK DEVICES 120
PLAYBACK MEDIA 86
PLAYSTATION 139, 140
PLL 57
PLUG 205
PLUG-AND-PLAY 11, 91, 93, 174
PLUG-IN KEYBOARD 147
PLUS SERVICE 70
POCKET LCD TV 35
POLARITY 74, 76
POLARIZED FEEDHORN 74
POLITICS 67
POLYGON GRAPHICS 134, 136
POPULAR ELECTRONICS 154
POPULAR MECHANICS 154
POPULAR SCIENCE 154
POPULUS 141
PORT 83
POST-PRODUCTION EDITING 120
POWER 75, 132, 143, 180
POWER ADAPTER 143
POWER AMPLIFIER 52
POWER MACINTOSH 120, 129
POWER OUTPUT 170
POWER SUPPLY 143
PRERECORDED FEATURE MOVIES 11
PRERECORDED DVD MOVIES 100
PRERECORDED MOVIE 12, 86, 95, 208
PRESENT 107
PREVIEW MONITORS 127
PRICE GUARANTEE 158
PRICING 79
PRIMARY COLOR 114
PRIMESTAR 77, 79, 80, 82, 83, 164
PRINTERS 147
PRO LOGIC 48, 49, 50, 163, 169, 175
PROBLEMS 205, 207
PROCESSING 124
PROCESSING POWER 12
PROCESSING SPEED 135
PROCESSOR 114, 132
PRODUCT INFORMATION 155
PROFESSIONAL 119
PROFESSIONAL QUALITY VIDEO 126
PROFESSIONAL VIDEOS 128
PROGRAM 83, 124
PROGRAMMABLE MEMORY REMOTE 60
PROGRAMMING 11, 70, 77, 83
PROGRAMMING ACCESS CARD 78
PROGRAMMING CIRCUITS 91
PROGRAMMING GUIDES 172
PROGRAMMING LANGUAGE 145
PROJECTION TV 8, 35, 38, 41
PROJECTION UNITS 38
PROMOTIONAL MATERIALS 167
PROPRIETARY CIRCUITRY 50
PROSCAN 80
PSYGNOSIS 141
PULSE CODE MODULATION STEREO
    (PCM) 114

PURCHASE 157
PURCHASE ADVICE 43
PURCHASER 153
PURCHASING 153
PURCHASING FRAUD 154, 155
PURCHASING HINTS 45
PURCHASING STRATEGIES 52, 157
PURCHASING TIPS 158
PUSH-ON MALE 181
PUZZLE GAMES 140

Q

Q-SOUND 50
Q-TIP 144
QUALITY 187
QUASI S-VHS PLAYBACK 174
QUICK-START 174

R

R.A. EDIT (RANDOM ASSEMBLE EDITING) 177
RABBIT EARS 7, 65
RADIO 14
RADIO FREQUENCY (RF) 181
RADIO FREQUENCY BAND 54
RADIO SHACK 65, 82, 207
RADIO WAVES 54, 64
RAM 125, 129, 145, 147
RANDOM PRESET MEMORY 170
RARE-EARTH ELEMENTS 35
RATE 54
RAW VIDEO FOOTAGE 119
RCA 67, 80, 100
RCA AUDIO JACK 56
RCA CINCH CORDS 208
RCA JACKS 16
RCA-TYPE CONNECTORS 184
RCA-TYPE PORTS 117
READING DEVICE 142
REAL-TIME COMMUNICATIONS 148
REAR PROJECTION TVS 35
REAR SPEAKERS 48
REAR SURROUND SPEAKERS 50
RECEIVER 10, 19, 48, 51, 52, 59, 66, 69, 76, 77, 79, 83
RECEIVER-TRANSMITTER 80
RECEIVING 65
RECHARGEABLE BATTERY PACKS 117
RECOMMENDATIONS 9, 10, 11, 12, 60, 93, 159
RECORD 97, 121, 143
RECORD BUTTON 87
RECORD PLAYER 98
RECORDABLE 100
RECORDABLE DISK 95
RECORDABLE UNIT 11
RECORDER 104
RECORDING 21, 86, 87, 110, 111
RECORDING DATE 113
RECORDING HEADS 107

RECORDING TECHNOLOGY 100
RECREATION ROOM 41
RED 28, 29, 32, 34, 35, 114
REEDER, CURT 94
REFLECTIONS 45
REFLECTIVE GHOSTING 64
REFLECTIVE WAVE SIGNAL 64
REFLECTOR 66
RELAY 71
REMOTE 11, 59
REMOTE CIRCUITS 91
REMOTE CONTROL 9, 52, 58, 68, 76, 78, 79, 91, 110, 112, 123, 124, 128, 144, 147, 170, 171, 177
REMOTE CONTROL FEATURES 91
REMOTE CONTROL FINDER 171
REMOTE DUPLICATION 59
REMOTE FINDER 60
REMOTE-CONTROLLED RECORDING 117
REPAIR BILL 207
REPAIR SHOP 205, 207
REPAIRS 94, 207
REPLACE 207
RERUN NETWORKS 70
RERUNS 67
RESEARCH 154
RESOLUTION 21, 27, 41, 83, 92, 96, 100, 108, 114, 149, 160, 161, 175
RETAIL PRICE 154
REVIEWS 154, 167
REWIND 173
RF 147
RF APPLICATIONS 184
RF AUDIO 196
RF BOX 142, 143, 147
RF BROADCAST SIGNALS 58
RF CABLE 55, 182
RF CONNECTORS 92, 181
RF LINES 187
RF SIGNAL 55, 65
RG58 184
RG59 184
RG6 184
RG8 184
RGB PICTURE TUBE 29
RIGHT STEREO CHANNEL 48
RISC PROCESSORS 135
ROLE PLAYING GAMES (RPGS) 140
ROM 132, 147
ROM CARTRIDGE 132
ROMANS 131
ROOF 81
ROOM 40
ROOM LIGHT 44
ROTATING DRUM 89
ROTATIONS-PER-MINUTE (RPM) 97
ROTOR SYSTEM 76
ROUND COAXIAL CABLE 184
RPG 140
RS-232 PORT 96, 114

RU-6U 74
RUBBING ALCOHOL 144

S

S-VHS 42
S-VIDEO 101
S-CONNECTION 92
S-INPUT 21
S-INPUT JACKS 162
S-VHS 92, 96, 105, 114
S-VHS CONNECT 21
S-VHS EDITING VCR 128
S-VHS TAPES 11
S-VHS-C 12, 92, 106, 108, 113, 116, 164
S-VIDEO 98, 147
S-VIDEO CABLES 185
S-VIDEO CONNECTORS 185
S-VIDEO OUTPUT 175
S-VIDEO PORTS 117
SALESPERSON 26, 153, 155
SAMSUNG 80, 100
SANYO 80, 157
SAP 46, 47, 51
SATELLITE 21, 23, 54, 58, 67, 71, 72, 73, 74, 75, 79, 80, 81, 83
SATELLITE ANTENNA 21, 73
SATELLITE CHANNELS 77
SATELLITE COMPANIES 85
SATELLITE DISH 6, 21, 63, 65, 67, 71, 72, 74, 79, 81, 210
SATELLITE NETWORKS 83
SATELLITE RECEIVER 19, 53
SATELLITE SERVICE 17
SATELLITE SIGNAL 54, 76
SATELLITE SYSTEMS 70, 77, 85
SATELLITE TV 63
SATELLITE UPLINK 72
SATURATION 30
SATURN 23, 137, 138, 139, 148
SATURN LOOP 65
SCAN 100
SCAN LINES 32, 33, 42, 43
SCART 92, 98
SCART CABLES 185
SCART CONNECTOR 180, 185
SCENES 121, 129
SCHMIDT OPTICAL SYSTEM 35
SCHMIDT PROJECTION TUBES 35
SCIENCE 70
SCRAMBLED SIGNALS 77
SCRAMBLING 68
SCREEN 8, 27, 28, 30, 34, 41, 60, 161
SCREEN FORMATS 37
SCREEN MEASUREMENT 35
SCREW ROD 76
SCREW TERMINALS 182
SCREW-ON MALE 181
SEARS 156
SECOND AUDIO PROGRAM 47
SECURITY 150

SEGA 134, 135, 136, 137, 138, 139, 140, 141, 148
SEGA CD-ROM UNIT 135
SEGA CHANNEL 136
SEGA GAME GEAR (GG) 134
SEGA GENESIS 23, 134, 141
SEGA MASTER SYSTEM (SMS) 134, 141
SEGA NOMAD 134
SEGA PICO 136
SEGA SATURN 135
SEGA SATURN NETLINK 148
SELECTOR DIALS 57
SELF-TIMER 177
SEMIAUTOMATIC FUNCTIONS 128
SEMICONDUCTOR 108
SEMICONDUCTOR CHIP 108
SEMIPROFESSIONAL 126
SENSORS 108
SEPARATE AUDIO PROGRAM (SAP) 169
SERVER 147
SERVICE 81
SERVICE HELP 157
SERVICE SHOP 206
SHADING 134
SHADOW MASK 33, 44
SHADOW MASK TECHNOLOGY 44
SHADOWS 126
SHARP 113, 116
SHARPNESS 45
SHARPNESS CONTROL 168
SHIELDED CABLES 184
SHUTOFF 173
SHUTTLE CONTROLS 174
SIDE-FIRING SPEAKERS 169
SIGHT 27, 28, 56, 54, 159
SIGNAL 21, 26, 42, 52, 53, 56, 57, 58, 59, 63, 68, 72, 73, 74, 76, 80, 95, 98, 99, 159, 180, 208
SIGNAL AMPLIFIERS 71, 188, 192
SIGNAL CHOICES 68
SIGNAL CIRCUITRY 26
SIGNAL DEGRADATION 80
SIGNAL GAIN 64, 192
SIGNAL NOISE 69
SIGNAL PATH 72
SIGNAL PICKS 68
SIGNAL PROBLEMS 208
SIGNAL QUALITY 27, 68
SIGNAL RECEIVER 63
SIGNAL SCRAMBLING 68
SIGNAL SELECTOR 198, 199
SIGNAL SOURCE 121
SIGNAL SPLITTERS 189, 179
SIGNAL STRENGTH 64
SIGNAL STRENGTH FEATURE 81
SIGNAL STRENGTH METER 81
SIGNAL TECHNOLOGY 27
SIGNAL TRANSMISSION 54
SIGNAL TRAPPING 68
SIGNAL TRAPS 69
SIGNAL-TO-NOISE RATIOS 106

SIGNALS 29,
    45, 51, 58, 76, 86, 90, 143, 185
SILICON GRAPHICS 134
SIMULATOR GAMES 140
SLEEP TIMER 170
SLOW-MOTION 174
SMALL SATELLITE SYSTEM 63
SMARTCARD 149
SMS 134, 140
SNAPPY 116
SNES 23, 140, 141
SNOW 67
SOAP OPERA 41
SOFTWARE 120, 124, 125, 128, 129, 132, 141, 147, 149
SOLID-STATE 107
SOLVENT 144
SONIC THE HEDGEHOG 138
SONY 59,
    80, 81, 87, 100, 104, 105, 113, 117, 147
SONY PLAYSTATION 23, 136, 137, 141
SONY WEBTV 148
SOUND 21, 27, 45,
    46, 48, 52, 54, 58, 61, 72, 83, 96,
    98, 101, 105, 112, 114, 133, 135, 141,
    149, 159, 162, 168
SOUND BITES 52
SOUND CHANNELS 47, 50
SOUND CIRCUITS 26, 45, 90
SOUND COMPONENTS 104
SOUND EFFECTS 49, 50, 129
SOUND EFFECTS MIXER 127
SOUND ENCOUNTER 52
SOUND EXPERIENCE 45
SOUND EXPLANATIONS 45
SOUND FIELD 45
SOUND FIELD PATTERNS 170
SOUND INFORMATION 23
SOUND OPTIONS 46
SOUND ROOMS 161
SOUND SIGNAL 95
SOUND SPACE 50
SOUND STAGE 47, 50
SOUND SYSTEM 10, 27, 52, 135, 136
SOUNDTRACKS 10, 89, 90, 119, 127, 129
SP (STANDARD PLAY) 105
SPACE 58, 114
SPEAKERS
    10, 16, 19, 21, 26, 27, 45, 46, 47, 48,
    49, 50, 51, 52, 72, 160, 163
SPEAKERS ON/OFF 169
SPECIAL EFFECTS 96, 98, 124, 129
SPECIAL EFFECTS A/V RECEIVER 159
SPECIALTY STORE 156
SPECTATOR 131
SPEED 54
SPIELBERG, STEVEN 119
SPIRAL TRACK 95
SPORTING EVENTS 69
SPORTS 70
SPORTS GAMES 140
SPOT BEAM 83

SQUARE FEET 50
SQUARESOFT 139, 140
SRS 50
SRS UNITS 52
STADIUM 131
STAINLESS STEEL 58
STAND-ALONE SET-TOP BOX 145
STAND-ALONE UNIT 142
STAND-ALONE VIDEO EDITING EQUIPMENT 126
STANDARD FEATURES 111
STANDARDS 27, 114
STAR WARS 45, 119
STARFOX 140
STARSIGHT 87, 164, 171, 174
STATION 57, 68, 143
STEADY IMAGES 117
STEEL 76, 180
STEREO 9, 19, 21,
    46, 55, 68, 89, 95, 143, 159, 163
STEREO DECODERS 87
STEREO EFFECT 50
STEREO HI-FI VCR 201
STEREO JACKS 10
STEREO MUSIC SYNTHESIZER 135
STEREO SIGNALS SOURCES 127
STEREO SOUND 45, 47
STEREO SOUND SYSTEM 134
STEREO SOURCE 45
STEREO SYSTEM 10
STEREO TV COMPATIBLE 71
STILL IMAGE 96, 114
STILL PICTURE 113
STORAGE SPACE 135
STORYBOARD 121, 129
STRATEGY GAMES 141
STREET FIGHTER 139
STUDENTS 119, 144
STUDIO MASTER 100
STYLING 38
STYLUS 98
STYLUS PAD 136
SUBSYSTEMS 104
SUBWOOFER 21, 50, 163, 169, 191
SULLIVAN, ED 27
SUN MICROSYSTEMS, INC. 145, 146, 149
SUPER GAMEBOY 133
SUPER LO-LUX 177
SUPER MARIO BROS 133, 138
SUPER NINTENDO ENTERTAINMENT SYSTEM (SNES) 133
SUPER STATIONS 70
SUPER VHS 92, 95
SUPER VHS COMPACT 106
SUPER-HIGH-SPEED REWIND 174
SUPER-VHS 11, 128
SUPERHET 57
SUPERHET RECEIVER 56
SUPERHETERODYNE (SUPERHET) RECEIVING SYSTEM 57
SUPPORT CIRCUITS 26
SUPPORT ELECTRONICS 104

SUPPORT EQUIPMENT 12
SUPPORT LINKAGES 88
SUPPORTING EQUIPMENT 104
SURFACE AREA 74
SURROUND 48
SURROUND CHANNEL 48
SURROUND SOUND 9, 10, 21, 46,
    47, 48, 50, 51, 52,
    68, 96, 101, 135, 159, 162, 163, 195
SURROUND SOUND DECODER 10, 27, 55
SURROUND SOUND RECEIVER 10
SURROUND SOUND SYSTEM 10
SURROUND SOUND TRICKS 51
SURROUND SOUND TV 51
SURROUND SPEAKERS 46, 169
SWEET SPOT 41
SYMPHONY 51
SYNC STABILIZER 208
SYNCHRONIZATION HEAD 88, 90
SYSTEM 41, 51, 61, 82, 138, 142, 149, 159
SYSTEM NOISE 47

T

T*HQ 139
TABLE OF CONTENTS 113
TABLETOP MODEL 40
TALK SHOW 41
TAPE 87, 88, 89, 90,
    105, 106, 107, 110, 121, 201
TAPE BACKUP UNIT 87
TAPE COUNTER 121, 174
TAPE INDICATORS 110
TAPE LENGTH 113
TAPE STANDARD 113
TAPE-REMAINING DISPLAY 176
TARGET 157
TECHNICAL SUPPORT STAFF 144
TECHNICIAN 205
TECHNOLOGY 26, 35, 46, 48, 80, 99, 145
TEKKEN 139
TELECOMMUNICATIONS ACT OF 1996 71
TELEVISION 7, 11, 26, 32,
    49, 59, 61, 64, 136, 145, 165, 167,
    196, 197, 198, 209
TELEVISION BROADCASTS 74
TELEVISION SCREEN 29
TELEVISION SET 25, 27, 45, 57, 161
TELEVISION TECHNOLOGY 61
TELEVISION TUBE 30
TERMINOLOGY 132
TERMS 64
TETRIS 140, 141
TEXTURE MAPPING 135
TEXTURED POLYGON GRAPHICS 134
TEXTURES 30, 137
THE MOVIE CHANNEL 68
THE RIGHT ANTENNA 67
THEATER 50
THEME 130
THREE-DIMENSIONAL GRAPHICS 134
THROUGH-HOLE SYSTEM 111

THROUGH-THE-LENS VIEWFINDER 111
THUMBS UP THUMBS DOWN 123
THX 10, 19, 45, 50, 52
THX CINEMA 10
THX HOME THEATER 170
THX PACKAGE 50
THX PROCESSING 50
THX-1138 118
TIME 93
TIME CODES 114
TIMERS 174
TIMING 56
TIMING 21, 54, 56, 88
TIMING INFORMATION 53, 89
TIMING SIGNAL 90
TINTS 30, 44
TITLE GENERATOR 174
TITLE MAKERS 126
TITLES 119, 126, 128, 177
TITLING 129
TOMB RAIDER 139
TONE CONTROLS 170
TONES 30
TOSHIBA 80, 100
TOTAL RECALL 37
TOWER LOCATIONS 65
TRACKING 205
TRADE MAGAZINES 154
TRANSDUCER 52
TRANSMITTER 59, 72
TRANSMITTING ANTENNAS 71
TRANSPONDERS 74, 75, 80
TRAPPING 68
TRIAD 28
TRIPOD 112, 117
TRISTAR 100
TROUBLESHOOTING GUIDE 208
TS2743C 59
TTL 111
TUBE 32, 33, 44, 161
TUBE SIZE 38
TUBE TECHNOLOGY 34
TUNER 55, 56, 57
TUNER CIRCUITS 91
TURBINE GENERATOR 107
TV 5, 10, 12, 14, 19, 41, 43, 44,
    45, 52, 55, 56, 57, 59, 60, 63,
    66, 72, 83, 86, 87, 90, 95, 98,
    107, 108, 110, 114,
    131, 143, 144, 146,
    147, 150, 151, 154, 159, 160, 162
TV BROADCASTS 63
TV CHANNEL 54, 57
TV DRIVE CIRCUITS 55
TV FORMAT 90
TV GUIDE 61, 70, 91
TV NEWS CREWS 106
TV PURCHASING CONSIDERATIONS 38
TV REMOTE 59
TV SCREEN 28, 32, 44, 134, 137, 143
TV SCREEN FORMATS 37
TV SCREENS 16

TV SET 7, 9, 10, 25, 27, 34, 40, 41, 43, 44, 50, 51, 57
TV SET-TOP BOXES 144
TV SHOWS 7, 63, 87
TV SIGNAL 17, 52, 53, 54, 55, 58, 64, 66, 67, 76
TV SPEAKERS 47, 49, 95
TV STATION 17, 64, 65, 66, 72
TV TECHNOLOGY 26, 38
TV TUBE SIZES 41
TV-TOP BOX 76, 83
TV-TOP CABLE BOXES 68
TV-TOP INDOOR ANTENNA 65
TV-TOP INTEGRATED RECEIVER/ DECODER (IRD) 78
TV/VCR COMBO 41
TWIN CABLE 55
TWIN-LEAD 182
TWO-DIMENSIONAL SURFACE 47
TWO-HEAD VCR 93, 164
TWO-VCR METHOD 121
TWO-VCR METHOD 120
TYPICAL DSS 79

U

U.S. 57, 68
UHF 64, 66, 183, 188
UHF-ONLY UNIT 65
ULTIMA 140
ULTIMATE ELECTRONICS 156
ULTRA HIGH FREQUENCY (UHF) 65
ULTRA-LOW NOISE 116
UNIDEN 80
UNITED STATES 20
UNITED STATES SATELLITE BROADCASTING 82
UNIVERSAL REMOTE 59, 60, 91, 161, 162
UNIX 145
UPGRADE 71, 147
UPLINK 75
UPLINK TRANSMITTERS 71
USER PROGRAMMABLE AUDIO 171
USER'S GUIDE 144
USSB 79, 82, 164

V

VACUUM 58
VACUUM TUBE 107
VARACTOR TUNERS 57
VARIABLE ATTENUATOR 64
VARIABLE FOCUS 109
VARIABLE-SPEED ZOOM 177
VCR 11, 12, 19, 41, 42, 55, 57, 59, 60, 61, 66, 68, 83, 85, 86, 87, 88, 91, 92, 93, 94, 95, 96, 101, 104, 105, 107, 108, 119, 120, 121, 123, 124, 126, 128, 129, 130, 143, 150, 154, 159, 161, 163, 165, 197, 198, 201, 205, 211

VCR 3/4 SWITCH 208
VCR COMPONENTS 107
VCR ERASE HEAD 107
VCR FILTER 70
VCR FORMAT 88
VCR HEADS 86
VCR PLUS+ 91, 161, 164, 175
VCR PLUS+ PROGRAMMER 60
VCR REPAIRS 94
VCR SIGNAL 12
VCR TAPES 63
VCR TIMER 70
VCR TUNERS 69
VCR USES 87
VCR/CAMCORDER SYMBIOSIS 128
VCR/TV COMBO 94
VCRPLUS+ 11, 87, 91, 93
VERTICAL 56
VERTICAL BLANKING PULSES 56
VERTICAL LINES 42
VERTICAL RESOLUTION 43
VERY HIGH FREQUENCY (VHF) 64, 65, 66, 182, 188
VHF APPLICATIONS 184
VHF/UHF COMBO UNIT 65
VHS 11, 87, 92, 105
VHS CAMCORDERS 104
VHS CASSETTE 105, 114
VHS HI-FI 92
VHS PICTURE 108
VHS TAPE 107
VHS VCR 164
VHS-C 12, 92, 105, 106, 113, 164
VHS-C CASSETTE 105
VHS-C FORMAT 104
VIDEO 5, 6, 12, 16, 21, 22, 23, 56, 58, 81, 86, 89, 101, 105, 124, 130, 142, 143, 191
VIDEO BOARDS 124
VIDEO CABLE 184
VIDEO CAMERA 12, 17, 103, 104, 107
VIDEO CAPTURE BOARD 120, 124, 130
VIDEO CAPTURE CARD 128
VIDEO CARD 125, 129
VIDEO CASSETTE PLAYERS (VCPS) 94, 121
VIDEO CASSETTE RECORDERS (VCRS) 86, 94, 172
VIDEO CD 100
VIDEO COMPONENTS 11, 17, 23
VIDEO COMPRESSION 114
VIDEO CONNECTORS 184
VIDEO DISK 95
VIDEO DISK MOVIES 97
VIDEO DISK PLAYER 85, 214
VIDEO DISKS 97
VIDEO DRUM 90
VIDEO DUBBING 87
VIDEO EDITING 11, 94, 114, 119, 124, 126
VIDEO EFFECTS 128
VIDEO EQUIPMENT 156, 208
VIDEO EQUIPMENT DEALERS 27

VIDEO FOOTAGE 128, 130
VIDEO GAME 98, 133, 143
VIDEO GAME CARTRIDGE 23
VIDEO GAME CONSOLE 12, 19, 23,
    131, 143, 145
VIDEO GAME CONSOLE UNIT 132
VIDEO GAME INDUSTRY 131
VIDEO GAME SYSTEM 132, 141
VIDEO GAME UNITS 23
VIDEO HARDWARE 179
VIDEO HEAD DRUM 89
VIDEO HEADS 89, 93, 161, 164, 107, 173
VIDEO IMAGES 95, 127
VIDEO INDUSTRY 27, 113
VIDEO INFORMATION 80, 114
VIDEO MIXERS 127, 128
VIDEO NOISE 116
VIDEO NOISE REDUCTION (VNR) 168
VIDEO ON DEMAND 69
VIDEO OUTPUTS 147, 172
VIDEO PALETTE 127
VIDEO PRESETTINGS 171
VIDEO PROCESSORS 126, 127, 128
VIDEO QUALITY 92
VIDEO RECORDING CAMERAS 103
VIDEO RECORDING DEVICES 7
VIDEO RELEASE 87
VIDEO RESOLUTION 114, 160
VIDEO SELECTORS 179, 189
VIDEO SIGNAL 23, 51, 55,
    56, 88, 114, 127, 142, 187
VIDEO STABILIZER 127
VIDEO STORES 97, 141
VIDEO SYSTEM 6, 25, 27, 55, 188
VIDEO TECHNOLOGY 6, 23, 60
VIDEO TERMS 7
VIDEO TOASTER SOFTWARE 129
VIDEO WORLD 12, 151
VIDEO-8 105
VIDEO-CIPHER II 77
VIDEO-ON-DEMAND 71
VIDEO-QUALITY SHIELDED CABLES 184
VIDEO/SOUND EFFECT MIXERS 126
VIDEOCASSETTE 205, 208
VIDEOCASSETTE RECORDER 17
VIDEONICS, INC. 123, 127
VIDEONICS THUMBS UP 2000 128
VIDEOTAPE 87, 89, 90, 91, 121, 126
VIEWER 25, 35
VIEWFINDER 104, 110, 111, 164
VIEWING 44
VIRTUA FIGHTER 139
VISUAL 27
VISUAL STIMULATION 27
VOLUME CONTROL 70
VOLUME LEVELING 169
VR4106 59

W

WALL OUTLETS 179
WALL PLUG 180

WALL SOCKET 143, 205
WALMART 157
WARNER 100
WARRANTY 206
WARRANTY PAPERS 144
WATTAGE 80, 107, 164
WATTS PER CHANNEL 163
WAVE 54
WAVELENGTH 54, 67
WAVES 54, 58
WEAR 96
WEB 145
WEB ADDRESS 151
WEB PAGES 144
WEB SURFING 83
WEBSITE 48, 80, 146, 150, 151, 155, 160,
    164, 167
WEBTV 147
WEBTV NETWORKS 147
WEBTV UNIT 147
WEIGHT 38, 116
WHITE DOT 28
WIDE ANGLE 109
WIDE MODE 177
WIDE-BAND DATA PORT 83
WIDESCREEN 8, 114
WIDESCREEN RATIO 114
WINDOWS 145
WINDOWS '95 129, 145
WINDOWS NT 129
WIPES 127, 128, 129, 176
WIRE 19, 26, 54, 90, 142, 143, 179, 180
WIRELESS CABLE 71
WIRELESS CIRCUITS 123
WORD PROCESSOR 145, 150
WORLD WAR II 26
WORLD WIDE WEB 19, 144
WWII 34
WWW BROWSER 148

X

X 109

Y

Y 30, 56
Y SIGNAL 30
Y/C 92
YAGI ANTENNAS 66
YOKE 32, 33

Z

ZENITH 59, 100
ZIP CODE 81
ZOOM 109

Howard W. Sams
A Bell Atlantic Company

Your Technology Connection to the Future!

Now You Can Visit Howard W. Sams & Company <u>On-Line</u>:
http://www.hwsams.com

Gain Easy Access to:

- The PROMPT Publications catalog, for information on our *Latest Book Releases*.
- The PHOTOFACT Annual Index.
- Information on Howard W. Sams' Latest Products.
- *AND MORE!*

PROMPT® PUBLICATIONS

CALL 1-800-428-7267 TODAY FOR THE NAME OF YOUR NEAREST PROMPT PUBLICATIONS DISTRIBUTOR

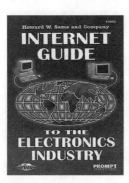

Internet Guide to the Electronics Industry
John Adams

Although the Internet pervades our lives, it would not have been possible without the growth of electronics. It is very fitting then that technical subjects, data sheets, parts houses, and of course manufacturers, are developing new and innovative ways to ride along the Information Superhighway. Whether it's programs that calculate Ohm's Law or a schematic of a satellite system, electronics hobbyists and technicians can find a wealth of knowledge and information on the Internet.

In fact, soon electronics hobbyists and professionals will be able to access on-line catalogs from manufacturers and distributors all over the world, and then order parts, schematics, and other merchandise without leaving home. The *Internet Guide to the Electronics Industry* serves mainly as a directory to the resources available to electronics professionals and hobbyists.

Internet
192 pages - Paperback - 5-1/2 x 8-1/2"
ISBN: 0-7906-1092-2 - Sams: 61092
$16.95 ($22.99 Canada) - December 1996

Real-World Interfacing with Your PC
James "J.J." Barbarello

As the computer becomes increasingly prevalent in society, its functions and applications continue to expand. Modern software allows users to do everything from balance a checkbook to create a family tree. Interfacing, however, is truly the wave of the future for those who want to use their computer for things other than manipulating text, data, and graphics.

Real-World Interfacing With Your PC provides all the information necessary to use a PC's parallel port as a gateway to electronic interfacing. In addition to hardware fundamentals, this book provides a basic understanding of how to write software to control hardware.

While the book is geared toward electronics hobbyists, it includes a chapter on project design and construction techniques, a checklist for easy reference, and a recommended inventory of starter electronic parts to which readers at every level can relate.

Computer Technology
119 pages - Paperback - 7-3/8 x 9-1/4"
ISBN: 0-7906-1078-7 - Sams: 61078
$16.95 - March 1996

CALL 1-800-428-7267 TODAY FOR THE NAME OF YOUR NEAREST PROMPT PUBLICATIONS DISTRIBUTOR

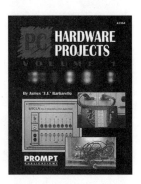

PC Hardware Projects Volume 1
James "J.J." Barbarello

Now you can create your own PC-based digital design workstation! Using commonly available components and standard construction techniques, you can build some key tools to troubleshoot digital circuits and test your printer, fax, modem, and other multiconductor cables.

This book will guide you through the construction of a channel logic analyzer, and a multipath continuity tester. You will also be able to combine the projects with an appropriate power supply and a prototyping solderless breadboard system into a single digital workstation interface!

PC Hardware Projects, Volume 1, guides you through every step of the construction process and shows you how to check your progress.

PROJECT SOFTWARE DISK INCLUDED!

Computer Technology
256 pages - Paperback - 7-3/8 x 9-1/4"
ISBN: 0-7906-1104-X - Sams: 61104
$24.95 - Feb. 1997

PC Hardware Projects Volume 2
James "J.J." Barbarello

PC Hardware Projects, Volume 2, discusses stepper motors, how they differ from conventional and servo motors, and how to control them. It investigates different methods to control stepper motors, and provides you with circuitry for a dedicated IC controller and a discrete component hardware controller.

Then, this book guides you through every step of constructing an automated, PC-controlled drilling machine. You'll then walk through an actual design layout, creating a PC design and board. Finally, you'll see how the drill data is determined from the layout and drill the PCB. With the help of the information and the data file disk included, you'll have transformed your PC into your very won PCB fabrication house!

PROJECT SOFTWARE DISK INCLUDED!

Computer Technology
256 pages - Paperback - 7-3/8 x 9-1/4"
ISBN: 0-7906-1109-0 - Sams: 61109
$24.95 - May 1997

CALL 1-800-428-7267 TODAY FOR THE NAME OF YOUR NEAREST PROMPT PUBLICATIONS DISTRIBUTOR

The Howard W. Sams Troubleshooting & Repair Guide to TV
Howard W. Sams & Company

The Howard W. Sams Troubleshooting & Repair Guide to TV is the most complete and up-to-date television repair book available. Included in its more than 300 pages is complete repair information for all makes of TVs, timesaving features that even the pros don't know, comprehensive basic electronics information, and extensive coverage of common TV symptoms.

This repair guide is completely illustrated with useful photos, schematics, graphs, and flowcharts. It covers audio, video, technician safety, test equipment, power supplies, picture-in-picture, and much more. *The Howard W. Sams Troubleshooting & Repair Guide to TV* was written, illustrated, and assembled by the engineers and technicians of Howard W. Sams & Company.

The In-Home VCR Mechanical Repair & Cleaning Guide
Curt Reeder

Like any machine that is used in the home or office, a VCR requires minimal service to keep it functioning well and for a long time. However, a technical or electrical engineering degree is not required to begin regular maintenance on a VCR. *The In-Home VCR Mechanical Repair & Cleaning Guide* shows readers the tricks and secrets of VCR maintenance using just a few small hand tools, such as tweezers and a power screwdriver.

This book is also geared toward entrepreneurs who may consider starting a new VCR service business of their own. The vast information contained in this guide gives a firm foundation on which to create a personal niche in this unique service business. This book is compiled from the most frequent VCR malfunctions Curt Reeder has encountered in the six years he has operated his in-home VCR repair and cleaning service.

Video Technology
384 pages - Paperback - 8-1/2 x 11"
ISBN: 0-7906-1077-9 - Sams: 61077
$29.95 ($39.95 Canada) - June 1996

Video Technology
222 pages - Paperback - 8-3/8 x 10-7/8"
ISBN: 0-7906-1076-0 - Sams: 61076
$19.95 ($26.99 Canada) - April 1996

CALL 1-800-428-7267 TODAY FOR THE NAME OF YOUR NEAREST PROMPT PUBLICATIONS DISTRIBUTOR

Desktop Digital Video
Ron Grebler

The Video Hacker's Handbook
Carl Bergquist

Desktop Digital Video is for those people who have a good understanding of personal computers and want to learn how video (and digital video) fits into the bigger picture. This book will introduce you to the essentials of video engineering, and to the intricacies and intimacies of digital technology. It examines the hardware involved, then explores the variety of different software applications and how to utilize them practically. Best of all, *Desktop Digital Video* will guide you through the development of your own customized digital video system. Topics covered include the video signal, digital video theory, digital video editing programs, hardware, digital video software and much more.

Geared toward electronic hobbyists and technicians interested in experimenting with the multiple facets of video technology, *The Video Hacker's Handbook* features projects never seen before in book form. Video theory and project information is presented in a practical and easy-to-understand fashion, allowing you to not only learn how video technology came to be so important in today's world, but also how to incorporate this knowledge into projects of your own design. In addition to the hands-on construction projects, the text covers existing video devices useful in this area of technology plus a little history surrounding television and video relay systems.

Video Technology
225 pages + Paperback + 7-3/8 x 9-1/4"
ISBN: 0-7906-1095-7 • Sams: 61095
$29.95 • June 1997

Video Technology
336 pages + Paperback + 7-3/8 x 9-1/4"
ISBN: 0-7906-1126-0 + Sams: 61126
$24.95 + September 1997

**CALL 1-800-428-7267 TODAY FOR THE NAME OF
YOUR NEAREST PROMPT PUBLICATIONS DISTRIBUTOR**

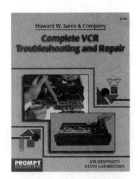

Howard W. Sams Complete VCR Troubleshooting & Repair
Joe Desposito & Kevin Garabedian

Complete VCR Troubleshooting and Repair contains sound VCR troubleshooting procedures beginning with an examination of the external parts of the VCR, then narrowing the view to gears, springs, pulleys, belts, and other mechanical parts. This book also shows how to troubleshoot tuner/demodulator circuits, audio and video circuits, display controls, servo systems, video heads, TV/VCR combination models, and more.

This book also contains nine VCR case studies, each focusing on a particular model of VCR with a specific problem. The case studies guide you through the repair from start to finish, using written instruction, helpful photographs, and Howard W. Sams' own *VCRfacts*® schematics.

Howard W. Sams Computer Monitor Troubleshooting & Repair
Joe Desposito & Kevin Garabedian

Computer Monitor Troubleshooting & Repair makes it easier for any technician, hobbyist or computer owner to successfully repair dysfunctional monitors. Learn the basics of computer monitors with chapters on tools and test equipment, monitor types, special procedures, how to find a problem and how to repair faults in the CRT. Other chapters show how to troubleshoot circuits such as power supply, high voltage, vertical, sync and video.

This book also contains six case studies which focus on a specific model of computer monitor. Using carefully written instructions and helpful photographs, the case studies guide you through the repair of a particular problem from start to finish. The problems addressed include a completely dead monitor, dysfunctional horizontal width control, bad resistors, dim display and more.

Video Technology
184 pages s Paperback s 8-1/2 x 11"
ISBN: 0-7906-1102-3 s Sams: 61102
$29.95 s March 1997

Troubleshooting & Repair
308 pages + Paperback + 8-1/2 x 11"
ISBN: 0-7906-1100-7 + Sams: 61100
$29.95 + July 1997

CALL 1-800-428-7267 TODAY FOR THE NAME OF YOUR NEAREST PROMPT PUBLICATIONS DISTRIBUTOR

Is This Thing On?
Gordon McComb

Advanced Speaker Designs
Ray Alden

Is This Thing On? takes readers through each step of selecting components, installing, adjusting, and maintaining a sound system for small meeting rooms, churches, lecture halls, public-address systems for schools or offices, or any other large room.

In easy-to-understand terms, drawings and illustrations, *Is This Thing On?* explains the exact procedures behind connections and troubleshooting diagnostics. With the help of this book, hobbyists and technicians can avoid problems that often occur while setting up sound systems for events and lectures.

Is This Thing On? covers basic components of sound systems, the science of acoustics, enclosed room, sound system specifications, wiring sound systems, and how to install wireless microphones, CD players, portable public-address systems, and more.

Advanced Speaker Designs shows the hobbyist and the experienced technician how to create high-quality speaker systems for the home, office, or auditorium. Every part of the system is covered in detail, from the driver and crossover network to the enclosure itself. Readers can build speaker systems from the parts lists and instructions provided, or they can actually learn to calculate design parameters, system responses, and component values with scientific calculators or PC software.

This book includes construction plans for seven complete systems, easy-to-understand instructions and illustrations, and chapters on sealed and vented enclosures. There is also emphasis placed on enhanced bass response, computer-aided speaker design, and driver parameters. *Advanced Speaker Designs* is a companion book to *Speakers for Your Home and Automobile*, also available from Prompt® Publications.

Audio Technology
136 pages - Paperback - 6 x 9"
ISBN: 0-7906-1081-7 - Sams: 61081
$14.95 ($20.95 Canada) - April 1996

Audio Technology
136 pages - Paperback - 6 x 9"
ISBN: 0-7906-1070-1 - Sams: 61070
$16.95 ($22.99 Canada) - July 1995

CALL 1-800-428-7267 TODAY FOR THE NAME OF YOUR NEAREST PROMPT PUBLICATIONS DISTRIBUTOR

Speakers for Your Home & Automobile
Gordon McComb, Alvis J. Evans, & Eric J. Evans

The cleanest CD sound, the quietest turntable, or the clearest FM signal are useless without a fine speaker system. This book not only tells readers how to build quality speaker systems, it also shows them what components to choose and why. The comprehensive coverage includes speakers, finishing touches, construction techniques, wiring speakers, and automotive sound systems.

Gordon McComb has written over 35 books and 1,000 magazine articles which have appeared in such publications as *Popular Science*, *Video*, *Omni*, *Popular Electronics*, and *PC World*. His writings has spanned a wide range of subjects, from computers, to video, to robots. Alvis and Eric Evans are the co-authors of many books and articles on the subject of electricity and electronics. Alvis is also an Associate Professor of Electronics at Tarrant County Junior College in Ft. Worth, Texas.

Audio Technology
164 pages - Paperback - 6 x 9"
ISBN: 0-7906-1025-6 - Sams: 61025
$14.95 ($20.95 Canada) - November 1992

Sound Systems for Your Automobile
Alvis J. Evans & Eric J. Evans

This book provides the average vehicle owner with the information and skills needed to install, upgrade, and design automotive sound systems. From terms and definitions straight up to performance objectives and cutting layouts, *Sound Systems* will show the reader how to build automotive sound systems that provide occupants with live performance reproductions that rival home audio systems.

Whether starting from scratch or upgrading, this book uses easy-to-follow steps to help readers plan their system, choose components and speakers, and install and interconnect them to achieve the best sound quality possible. Installations on specific types of vehicles are discussed, including separate chapters on coupes and sedans, hatchbacks, pickup trucks, sport utility vehicles, and vans.

Audio Technology
124 pages - Paperback - 6 x 9"
ISBN: 0-7906-1046-9 - Sams: 61046
$16.95 ($22.99 Canada) - January 1994

CALL 1-800-428-7267 TODAY FOR THE NAME OF YOUR NEAREST PROMPT PUBLICATIONS DISTRIBUTOR

IC Projects: Fun for the Electronics Hobbyist & Technician
Carl Bergquist

    IC Projects was written for electronics hobbyists and technicians who are interested in projects for integrated circuits — projects that you will be able to breadboard quickly and test easily. Designed to be fun and user-friendly, all of the projects in this book employ integrated circuits and transistors.
    IC projects presents the experienced electronics enthusiast with instructions on how to construct such interesting projects as a LED VU meter, infrared circuit, digital clock, digital stopwatch, digital thermometer, electronic "bug" detector, sound effects generator, laser pointer, and much more!

TV Video Systems
L.W. Pena & Brent A. Pena

    Knowing which video programming source to choose, and knowing what to do with it once you have it, can seem overwhelming. Covering standard hard-wired cable, large-dish satellite systems, and DSS, *TV Video Systems* explains the different systems, how they are installed, their advantages and disadvantages, and how to troubleshoot problems. This book presents easy-to-understand information and illustrations covering installation instructions, home options, apartment options, detecting and repairing problems, and more. The in-depth chapters guide you through your TV video project to a successful conclusion.

Electronic Projects
256 pages ♦ Paperback ♦ 6 x 9"
ISBN: 0-7906-1116-3 ♦ Sams: 61116
$21.95 ♦ May 1997

Video Technology
124 pages - Paperback - 6 x 9"
ISBN: 0-7906-1082-5 - Sams: 61082
$14.95 ($20.95 Canada) - June 1996

CALL 1-800-428-7267 TODAY FOR THE NAME OF YOUR NEAREST PROMPT PUBLICATIONS DISTRIBUTOR

Theory & Design of Loudspeaker Enclosures
Dr. J. Ernest Benson

Making Sense of Sound
Alvis J. Evans

The design of loudspeaker enclosures, particularly vented enclosures, has been a subject of continuing interest since 1930. Since that time, a wide range of interests surrounding loudspeaker enclosures have sprung up that grapple with the various aspects of the subject, especially design. *Theory & Design of Loudspeaker Enclosures* lays the groundwork for readers who want to understand the general functions of loudspeaker enclosure systems and eventually experiment with their own design.

Written for design engineers and technicians, students and intermediate-to-advanced level acoustics enthusiasts, this book presents a general theory of loudspeaker enclosure systems. Full of illustrated and numerical examples, this book examines diverse developments in enclosure design, and studies the various types of enclosures as well as varying parameter values and performance optimization.

This book deals with the subject of sound — how it is detected and processed using electronics in equipment that spans the full spectrum of consumer electronics. It concentrates on explaining basic concepts and fundamentals to provide easy-to-understand information, yet it contains enough detail to be of high interest to the serious practitioner. Discussion begins with how sound propagates and common sound characteristics, before moving on to the more advanced concepts of amplification and distortion. *Making Sense of Sound* was designed to cover a broad scope, yet in enough detail to be a useful reference for readers at every level.

Audio Technology
244 pages - Paperback - 6 x 9"
ISBN: 0-7906-1093-0 - Sams: 61093
$19.95 ($26.99 Canada) - August 1996

Audio Technology
112 pages - Paperback - 6 x 9"
ISBN: 0-7906-1026-4 - Sams: 61026
$10.95 ($14.95 Canada) - November 1992

CALL 1-800-428-7267 TODAY FOR THE NAME OF YOUR NEAREST PROMPT PUBLICATIONS DISTRIBUTOR

Home Security Projects
Robert Gaffigan

Security Systems for Your Home & Automobile
by Gordon McComb

Home Security Projects presents the reader with many projects about home security, safety and nuisance elimination that can easily be built in the reader's own home for less than it would cost to buy these items ready-made. Readers will be able to construct devices that will allow them to protect family members and electrical appliances from mishaps and accidents in the home, and protect their homes and belongings from theft and vandalism.

This book shows the reader how to construct the many useful projects, including a portable CO detector, trailer hitch alignment device, antenna saver, pool alarm, dog bark inhibitor, and an early warning alarm system. These projects are relatively easy to make and the intent of *Home Security Projects* is to provide enough information to allow you to customize them.

Security Systems is about making homes safer places to live and protecting cars from vandals and thieves. It is not only a buyer's guide to help readers select the right kind of alarm system for their home and auto, it also shows them how to install the various components. Learning to design, install, and use alarm systems saves a great deal of money, but it also allows people to learn the ins and outs of the system so that it can be used more effectively. This book is divided into eight chapters, including home security basics, warning devices, sensors, control units, remote paging automotive systems, and case histories.

Gordon McComb has written over 35 books and 1,000 magazine articles which have appeared in such publications as *Omni* and *PC World*. In addition, he is the coauthor of PROMPT® Publication's *Speakers for Your Home and Auto*.

Projects
256 pages + Paperback + 6 x 9"
ISBN: 0-7906-1113-9 + Sams: 61113
$19.95 + September 1997

Projects
130 pages + Paperback + 6 x 9"
ISBN: 0-7906-1054-X + Sams: 61054
$16.95 + July 1994

CALL 1-800-428-7267 TODAY FOR THE NAME OF YOUR NEAREST PROMPT PUBLICATIONS DISTRIBUTOR

ES&T Presents TV Troubleshooting & Repair
Electronic Servicing & Technology Magazine

TV set servicing has never been easy. The service manager, service technician, and electronics hobbyist need timely, insightful information in order to locate the correct service literature, make a quick diagnosis, obtain the correct replacement components, complete the repair, and get the TV back to the owner.

ES&T Presents TV Troubleshooting & Repair presents information that will make it possible for technicians and electronics hobbyists to service TVs faster, more efficiently, and more economically, thus making it more likely that customers will choose not to discard their faulty products, but to have them restored to service by a trained, competent professional.

Originally published in *Electronic Servicing & Technology*, the chapters in this book are articles written by professional technicians, most of whom service TV sets every day.

Video Technology
226 pages - Paperback - 6 x 9"
ISBN: 0-7906-1086-8 - Sams: 61086
$18.95 ($25.95 Canada) - August 1996

ES&T Presents Computer Troubleshooting & Repair
Electronic Servicing & Technology

ES&T is the nation's most popular magazine for professionals who service consumer electronics equipment. PROMPT® Publications, a rising star in the technical publishing business, is combining its publishing expertise with the experience and knowledge of *ES&T's* best writers to produce a new line of troubleshooting and repair books for the electronics market. Compiled from articles and prefaced by the editor in chief, Nils Conrad Persson, these books provide valuable, hands-on information for anyone interested in electronics and product repair.

Computer Troubleshooting & Repair is the second book in the series and features information on repairing Macintosh computers, a CD-ROM primer, and a color monitor. Also included are hard drive troubleshooting and repair tips, computer diagnostic software, networking basics, preventative maintenance for computers, upgrading, and much more.

Computer Technology
288 pages - Paperback - 6 x 9"
ISBN: 0-7906-1087-6 - Sams: 61087
$18.95 ($26.50 Canada) - February 1997

CALL 1-800-428-7267 TODAY FOR THE NAME OF YOUR NEAREST PROMPT PUBLICATIONS DISTRIBUTOR

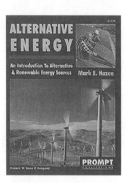

Alternative Energy
Mark E. Hazen

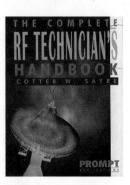

The Complete RF Technician's Handbook
Cotter W. Sayre

This book is designed to introduce readers to the many different forms of energy mankind has learned to put to use. Generally, energy sources are harnessed for the purpose of producing electricity. This process relies on transducers to transform energy from one form into another. *Alternative Energy* will not only address transducers and the five most common sources of energy that can be converted to electricity, it will also explore solar energy, the harnessing of the wind for energy, geothermal energy, and nuclear energy.

This book is designed to be an introduction to energy and alternate sources of electricity. Each of the nine chapters are followed by questions to test comprehension, making it ideal for students and teachers alike. In addition, listings of World Wide Web sites are included so that readers can learn more about alternative energy and the organizations devoted to it.

The *Complete RF Technician's Handbook* will furnish the working technician or student with a solid grounding in the latest methods and circuits employed in today's RF communications gear. It will also give readers the ability to test and troubleshoot transmitters, transceivers, and receivers with absolute confidence. Some of the topics covered include reactance, phase angle, logarithms, diodes, passive filters, amplifiers, and distortion. Various multiplexing methods and data, satellite, spread spectrum, cellular, and microwave communication technologies are discussed.

Cotter W. Sayre is an electronics design engineer with Goldstar Development, Inc., in Lake Elsinore, California. He is a graduate of Los Angeles Pierce College and is certified by the National Association of Radio and Telecommunications Engineers, as well as the International Society of Electronics Technicians.

Professional Reference
320 pages - Paperback - 7-3/8 x 9-1/4"
ISBN: 0-7906-1079-5 - Sams: 61079
$18.95 ($25.95 Canada) - October 1996

Professional Reference
281 pages - Paperback - 8-1/2 x 11"
ISBN: 0-7906-1085-X - Sams: 61085
$24.95 ($33.95 Canada) - July 1996

**CALL 1-800-428-7267 TODAY FOR THE NAME OF
YOUR NEAREST PROMPT PUBLICATIONS DISTRIBUTOR**